Microbes and Oral Squamous Cell Carcinoma

Samapika Routray
Editor

Microbes and Oral Squamous Cell Carcinoma

A Network Spanning Infection and Inflammation

Springer

Editor
Samapika Routray
Department of Dentistry
All India Institute of Medical Sciences
Bhubaneswar, Odisha, India

ISBN 978-981-19-0591-9 ISBN 978-981-19-0592-6 (eBook)
https://doi.org/10.1007/978-981-19-0592-6

This Springer imprint is published by the registered company Springer Nature Singapore Pte Ltd.
The registered company address is: 152 Beach Road, #21-01/04 Gateway East, Singapore 189721, Singapore

"In loving memory of Mrs Usha Misra"

Foreword

Every external and internal surface of a mammalian body carries a large and distinctive microbiome, varying by anatomical subsite, age, sex, ethnicity, diet, health, and disease. Microorganisms are fundamental to life, health, and happiness.

From very small to individually visible to the naked eye, these organisms are from all the three domains of living things on earth (and who knows what space exploration will reveal!). The Linnaean system of classification has stood the test of time. Our microbiota come from the domains of Bacteria, Archaea, and Eukarya. These life forms are driven by their DNA, contained within a nucleus in Eukaryotes and Archaea, nonnucleated in Bacteria. Then there are the obligate intracellular parasites, phages, and viruses, which infect cells of bacteria and eukaryotic organisms: they reproduce by hijacking chromosomal DNA of the host cell and can be constructed from DNA or from RNA only.

Underpinning all, classifications are increasingly based on DNA or genetic information.

The "load" of these microorganisms on and in our bodies substantially exceeds that of the total cellular components of our bodies. For much of the past century the ratio of number of nonhuman cells to constituent cells of our bodies has been thought to be orders of magnitude greater: this however has been revised down recently to numbers closer to 10:1 or equality. What is certain is that the genetic load—the numbers of unique genes—of our microbiota certainly exceeds our own by a large margin. Most of our microbes consort with each other and with our constituent cells—in both health and disease.

In studying associations of microbes with diseased tissues and organs in animals (and these principles apply to plants too), we have to consider the activity of consortia, of complex cooperative mixtures of viruses, bacteriophages, bacteria, archaea, fungi, and protea. Many diseases of mankind are indeed highly specific, caused by a single pathogenic species and specific strain thereof: e.g., AIDS by HIV and diphtheria by *Corynebacterium diphtheriae*. On the other hand, most common microbial afflictions are related to consortia: it is the total metabolic activity of the consortium, modified by host responses, which determine disease outcomes. This latter is true of common afflictions such as dental caries, periodontal diseases, gastroenteritis, and most cancers. Understanding the specificity of—or lack of—the role of the microbiota in disease is a huge challenge, now making fast progress because of advances in the technologies of molecular genetics.

The technologies produce huge amounts of data. It is difficult to see the wood for the trees. Unravelling the role of microbiota in the etiology, pathogenesis, and progression of oral cancer is a philosophical challenge. We are at the beginning of our understanding. This book is a brave step in the right direction. It is written predominantly by basic and clinical scientists from the Indian subcontinent and by lovers thereof from elsewhere. This is apt because the area has, and has had for millennia, the highest incidence and mortality rates from oral cancer in the world.

My final point: As scientists we are thrilled by the tools we have and the opportunities for collaborative research. But we do know the major causes of oral cancer and while we delve deeper and deeper into mechanisms, the world is failing in primary prevention of oral cancer (indeed of most communicable and noncommunicable diseases). We must play our role in health promotion by public education and contributions to good public health policies. Oral cancer is a preventable disease.

Newell W. Johnson

Menzies Health Institute
Queensland and School of Medicine
and Dentistry, Griffith University
Brisbane, QLD, Australia

Faculty of Dentistry, Oral and Craniofacial
Sciences, King's College London
London, UK
28 April 2022

Preface

Our human body (the host) and microbes have evolved and existed in a symbiotic relationship for millions of years. They together form a complex "superorganism" which benefits crucial aspects of life. Disruptions in the symbiotic relationship tends to promote disease. It has been hypothesized that this imbalance can initiate carcinomas of various organs, orchestrated by inflammatory response of the host, resulting in fabrication of carcinogenic metabolic products, or by the production of toxins, which disrupt the cell cycle. It is understood that the role of infection and chronic inflammation has turned out to be portentous in cancer development. It has become clear that infection, inflammation, and certain cancers are intimately linked, and interactions between pathogens and tumours are crucial. In recent years, inflammation is linked to pathogen-associated molecular patterns (PAMPs) and danger (or damage)-associated molecular patterns (DAMPs). Viruses are infamous to induce cell transformation, with unexpected connections between innate immunity, immune sensors, and tumour suppressor signalling. Similarly, bacteria such as *Helicobacter pylori* and *Salmonella enterica* subsp. *enterica* serovar Typhi, can promote different cancer types.

Oral squamous cell carcinoma (OSCC) continues to be a major global health problem. OSCC emerges from a multistep pathogenesis associated with cancer as summarized by Hanahan and Weinberg in "Hallmarks of Cancer". The modifying risk factors by which oral flora might cause OSCC are hypothetical like, e.g., release of carcinogenic substances, such as nitrosamine, conversion of ethanol to acetaldehyde, or release of pro-inflammatory mediators that can dysregulate cellular cycling, disrupt signalling mechanisms via bacterial toxins, and act as tumour promoters. Oral microorganisms inevitably upregulate inflammatory mediators that affect the complex metabolic pathways and may thus be implicated in the process of carcinogenesis. In dysbiosis, in particular, the precised equilibrium of the oral ecosystem is disrupted, allowing disease-promoting bacteria to manifest a compositional shift in the microbiota.

This book is a sincere attempt to summarize current concepts regarding the association between microorganisms and oral carcinogenesis and the possible oncogenic mechanisms enhancing it. Establishing the association of the oral microbiome with oral cancer risk may lead to significant advances in understanding of cancer aetiology, potentially opening a new research paradigm for cancer prevention.

Bhubaneswar, Odisha, India Samapika Routray

Contents

About the Editor

Samapika Routray, MDS, PhD works as Associate Professor at the Department of Dentistry, All India Institute of Medical Sciences, Bhubaneswar. She is a specialist in Oral Pathology & Microbiology. Her main research interest is in early detection of oral cancer using noninvasive modalities. She has published numerous papers in national and international high indexed journals and has authored multiple book chapters too. Her recent paper on "An integrated approach for identification of a panel of candidate genes arbitrated for invasion and metastasis of oral squamous cell carcinoma" using a multiomics approach is well appreciated.

Epidemiology of Oral Squamous Cell Carcinoma in Indian Scenario

1

Suzanne Tanya Nethan, Priyanka Ravi, and Prakash C. Gupta

Abstract

Oral squamous cell carcinoma (OSCC), comprising of more than 90% of all the malignant neoplasms of the oral cavity, is the 17th most common cancer globally. In India, it is the third most common cancer overall (both genders combined) but the most common cancer among males, and also ranks third among the various cancer types contributing to mortality in the country. The current chapter provides information on the various aspects of epidemiology of oral cancer/OSCC such as its incidence (including topographical), mortality, demography, and trends, in the global and Indian context, with detailed discussion about the latter.

Keywords

Epidemiology · India · Incidence · Mortality · Demography · Trends

S. T. Nethan (✉)
School of Preventive Oncology, Patna, Bihar, India
e-mail: suzanne.nethan@gmail.com

P. Ravi
Mel and Enid Zuckerman College of Public Health, University of Arizona, Tucson, Arizona, USA
e-mail: priyankaravi@email.arizona.edu

P. C. Gupta
Healis Sekhsaria Institute for Public Health, Navi Mumbai, Maharashtra, India
e-mail: pcgupta@healis.org

S. Routray (ed.), *Microbes and Oral Squamous Cell Carcinoma*,
https://doi.org/10.1007/978-981-19-0592-6_1

1.1 Introduction

Oral squamous cell carcinoma (OSCC) is the most common malignancy of the oral cavity, which represents up to 80–90% of all oral malignant neoplasms [1]. Tobacco is the primary risk factor for oral cancer; however, this condition may also be associated with various other behavioural risk factors such as areca nut and alcohol consumption, infectious risk factors such as human papilloma virus, and lack of protective factors in the form of inadequate dietary nutrition, to name a few [2]. The clinical definition of 'oral cancer' as suggested by the American Joint Committee on Cancer and the Union for International Cancer Control includes OSCC originating from the mucosa of the lips, cheeks, floor of the mouth, hard palate, alveolus, gingiva, retromolar trigone, and the anterior two-thirds of the tongue.

The current chapter provides information on the various aspects of epidemiology of oral cancer/OSCC in the global and Indian context, with detailed discussion about the latter.

1.2 Global Incidence and Mortality of Oral cancer

Lip and oral cavity cancer represents the 17th most common neoplasm worldwide, with almost 377,713 new cases and over 177,757 deaths estimated in 2020 [3]. Globally, the incidence and mortality rates (age-standardised) are consistently higher among males than females, the annual incidence rate is 6.0 per 100,000 among males and 2.3 among females, while the mortality rate is 2.8 among males and 1.0 among females [3].

According to the regional distribution, the highest overall incidence rates were noted in Melanesia (16.7), South-Central Asia (9.0), and Australia and New Zealand (6.0), ranging up to 22.2 in males and up to 11.9 in females. With regard to mortality, the highest overall mortality rates were noted in Melanesia (6.5), South-Central Asia (5.1), and Central and Eastern Europe (2.4) [4].

At the national level, in 2020 the highest age-standardised incidence rate of cancer of the oral cavity was reported in Papua New Guinea (21.2), followed by Pakistan (10.1), India (9.8), Sri Lanka (9.7), and Bangladesh (9.5). With regard to mortality, the highest age-standardised mortality rate due to oral cancer in 2020 was reported in Papua New Guinea (8.3) followed by Pakistan (6.4), Bangladesh (5.6), India (5.4), and Sri Lanka (4.5). With respect to gender variation, the highest age-standardised incidence and mortality rates were noted in Papua New Guinea for both genders, i.e. incidence rates: Males, $M = 28.4$, Females, $F = 14.8$; mortality rates: $M = 11.9$, $F = 5.1$ [3].

1.3 Oral Cancer Incidence and Burden in India

Oral cancer is the third common cancer overall (both genders combined) in India, with 135,929 new cases and 75,290 deaths reported in 2020 (Fig. 1.1). However, it is most common among males with an age-standardised incidence rate of 14.8

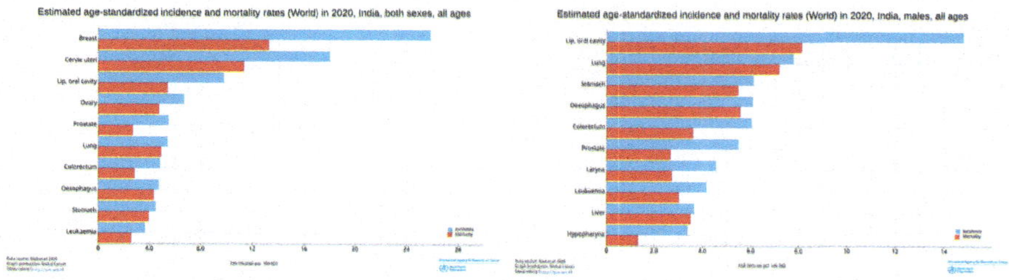

Fig. 1.1 Age-standardised incidence and mortality rates of oral cancer in India

(Fig. 1.1) as opposed to 4.6 among females; the observed mortality rate is 8.1 among males and 2.7 among females [3].

1.4 Survival and Mortality

In 2016, cancer of lip and oral cavity are among the top ten cancers responsible for the highest proportion of cancer disability-adjusted life-years (DALYs) in India, and it is the first or second leading cause of cancer deaths among Indian males [5].

Survival is found to be steadily declining with advanced age and advanced clinical stages [6]. The overall survival and disease-free survival was found to be better in patients without lymph node involvement and those patients treated in the early stage of cancer [7]. Tongue, buccal mucosa, and retromolar trigone cancers have been reported to have poor survival rates compared to lips [6].

In a 15-year follow-up of the oral cancer screening participants in Kerala, there was 47% reduction and 81% reduction in oral cancer mortality among those who complied with three and four rounds of oral cancer screening [8]. A retrospective study of 136 male oral cancer patients who have undergone surgery from June 2009 to June 2013 in rural Kerala shows 5-year survival rate of 79% among node-negative patients and 59% among the node-positive patients, and the median disease-free survival duration was 48.2 months [7]. The highest DALY for lip and oral cavity cancer was in Madhya Pradesh, followed by Gujarat and Kerala [5]. In a study conducted in Mumbai, the 5-year observed survival was 59.1% for localised cancer, 15.7% for cancers with regional extension, and 1.6% for those with distant metastasis [6].

1.5 Demography

1.5.1 Gender

Oral cancer is the most common type of cancer among males [5, 9], and the crude incidence rate of this type cancer was substantially higher in males than in females in India [5]. Among both males and females, mouth cancer had maximum age adjusted incidence rates (AARs) (64.8) in the central zone of India, and males observed maximum AAR tonsil cancer (36.3%), while females showed maximum AAR salivary gland cancer (14%) [10].

1.5.2 Age

Oral cancer is reported to be prevalent in the age group of 60–70 years in North India [9]. AAR of mouth and tongue cancers among males show cases at a very early age group of 0–20 years specially in the western region PBCRs followed by the north region PBCRs. Among (AAR) other cancers of associated sites around oral cavity, tonsil cancer shows cases at a very early period in life (0–9 years) in the north region PBCRs [10].

1.5.3 Socioeconomic Status

In a hospital-based study conducted in Tamil Nadu, it was observed that majority participants were from lower socioeconomic classes, and most rural participants had agriculture as occupation and most urban subjects were unskilled labourers [11]. The majority family income was below 5000 among both rural (94.9%) and urban (71.9%), and difference in the prevalence of oral cancer among different levels of literacy and occupation was found to be significant statistically [11]. Disadvantageous childhood socioeconomic position was found to be critical for oral cancer risk in Kerala [12].

1.5.4 Race and Ethnicity

The oral cancer incidence was found to be high in the central region of India, and north-eastern states showed higher oral cancer incidence among females [10]. The lower-middle epidemiological transition level (ETL) states which includes Arunachal Pradesh, Mizoram, Nagaland, Uttarakhand, Gujarat, Tripura, Sikkim, and Manipur had the highest crude incidence and DALY of lip and oral cavity cancer [5]. The difference in the genetic factors, behavioural, lifestyle, and food habits could be associated with the very peculiar type of cancer incidence pattern in the north-eastern states compared to mainland India [10].

1.6 Trends

The age-specific incidence rates of mouth cancer among men increased over the 25-year period [13]. In a tertiary care hospital in Mumbai, male to female ratio increased from 2.14 to 4.28 for tongue cancer and 2.69–4.59 for buccal mucosa cancer during the period of 1996–2012 [14]. The age-standardised incidence rate of lip and oral cavity decreased significantly by 6.4% from 1990 to 2016 in India and it varied 5.1 times among both sexes combined across the states [5]. An increased trend of 0.35% tongue and 0.6% buccal mucosa cancers were found a 16-year period in a tertiary care hospital in Mumbai [14].

1.7 Topographical Description/Incidence by ICD-10 Sub-Sites

The most recently published report from the National Cancer Registry Programme contains data on various cancers including cancers of the head and neck region (i.e. lip, tongue, mouth, tonsil, oropharynx, nasopharynx, hypopharynx, larynx) from 58 Hospital Based Cancer Registries (HBCRs) and 28 Population Based Cancer Registries (PBCRs) across India [15].

As reported by the PBCRs, mouth cancer contributed to one-third of the total head and neck cancers. It was also the highest contributor both in males ($n = 16{,}420$) and females ($n = 6301$). Among females, tongue cancer was the second highest contributor ($n = 3015$). In addition, data has also been provided regarding clinical extent of spread (localised, locoregional, distant metastasis, or unknown) of these cancers and the type of treatment (surgery, radiotherapy, systemic therapy, multimodality, palliative care) based on the same. The commonest clinical presentation of the extent of spread for both cancers of the tongue ($M = 58.6\%$, $F = 58.3\%$) and mouth ($M = 65.8\%$, $F = 70.2\%$) was locoregional. Based on the extent of disease, multimodality treatment was provided for tongue cancer among both males and females with locoregional and distant metastatic spread, while surgery was preferred among females with localised cancer. Multimodality treatment was also preferred for mouth cancer in both males and females, with surgery and radiotherapy as the second choice, based on the clinical extent of disease [15].

1.8 Regional Considerations

A significant variation has been noted in the regional distribution of oral cancer among males and females in India. East Khasi Hills district in Meghalaya (12.8 per 100,000) reported the highest age adjusted incidence rate (AAR) among males for tongue cancer followed by Ahmedabad urban in Gujarat (10.5), while among females, Bhopal in Madhya Pradesh (4.0) reported the highest AAR of tongue cancer followed by Cachar district in Assam (3.8). With regard to mouth cancer, Ahmedabad urban (19.5) had the highest AAR among males followed by Bhopal (15.9), while East Khasi Hills district (9.5) had the highest AAR of mouth cancer among females. As per the HBCR data, the proportion of lip cancer was highest in Western India ($n = 229$) among males while among females, it was highest in South India ($n = 131$) [15].

1.9 Conclusion

Oral cancer is among the most common cancers in India, which accounts for over 30% of all cancers reported in the country. Restricting the use of tobacco [16, 17], areca nut [18], and alcohol [19] which are recognised risk factors for oral cancers and earlier detection of oral cancers would reduce oral cancer deaths in India.

References

1. Johnson NW, Jayasekara P, Amarasinghe AAHK (2011) Squamous cell carcinoma and precursor lesions of the oral cavity: epidemiology and aetiology. Periodontol 57(1):19–37
2. Warnakulasuriya S (2009) Causes of oral cancer--an appraisal of controversies. Br Dent J 207(10):471–475
3. WHO. IARC (n.d.) Cancer Today. https://gco.iarc.fr/today/home
4. GCO (2020) Lip, oral cavity. Fact sheet. Globocan 2020. https://gco.iarc.fr/today/data/factsheets/cancers/1-Lip-oral-cavity-fact-sheet.pdf
5. India State-Level Disease Burden Initiative Cancer Collaborators (2018) The burden of cancers and their variations across the states of India: the Global Burden of Disease Study 1990-2016. Lancet Oncol 19(10):1289–1306
6. Yeole BB, Ramanakumar AV, Sankaranarayanan R (2003) Survival from oral cancer in Mumbai (Bombay), India. Cancer Causes Control 14(10):945–952
7. Thavarool SB, Muttath G, Nayanar S, Duraisamy K, Bhat P, Shringarpure K et al (2019) Improved survival among oral cancer patients: findings from a retrospective study at a tertiary care cancer centre in rural Kerala, India. World J Surg Oncol 17(1):15
8. Sankaranarayanan R, Ramadas K, Thara S, Muwonge R, Thomas G, Anju G et al (2013) Long term effect of visual screening on oral cancer incidence and mortality in a randomized trial in Kerala, India. Oral Oncol 49(4):314–321
9. Sahu PK, Kumar S (2019) Epidemiological aspects of oral cancer in North Indian population. Indian J Otolaryngol Head Neck Surg 71(Suppl 1):944–948
10. Sharma S, Satyanarayana L, Asthana S, Shivalingesh KK, Goutham BS, Ramachandra S (2018) Oral cancer statistics in India on the basis of first report of 29 population-based cancer registries. J Oral Maxillofac Pathol 22(1):18–26
11. Ganesh R, John J, Saravanan S (2013) Socio demographic profile of oral cancer patients residing in Tamil Nadu - a hospital based study. Indian J Cancer 50(1):9
12. ThekkePurakkal AS, Naimi AI, Madathil SA, Kumamangalam Puthiyannal SH, Netuveli G, Sacker A et al (2018) Differential impact of socioeconomic position across life on oral cancer risk in Kerala, India: an investigation of life-course models under a time-varying framework. Community Dent Oral Epidemiol 46(6):592–600
13. Gupta PC, Ray CS, Murti PR, Sinha DN (2014) Rising incidence of oral cancer in Ahmedabad city. Indian J Cancer 51(Suppl 1):S67–S72
14. Malik A, Mishra A, Garg A, Shetty R, Mair M, Chakrabarti S et al (2018) Trends of oral cancer with regard to age, gender, and subsite over 16 years at a tertiary cancer center in India. Indian J Med Paediatr Oncol 39(3):297
15. ICMR-NCDIR (2020) Report of National Cancer Registry Programme (ICMR-NCDIR), Bengaluru, India. https://www.ncdirindia.org/All_Reports/Report_2020/default.aspx

16. IARC (2004) Tobacco smoke and involuntary smoking. In: Monographs on the evaluation of carcinogenic risks to humans, vol 83. IARC, Lyon
17. IARC (2004) Smokeless tobacco and some tobacco-specific n-nitrosamines. In: Monographs on the evaluation of carcinogenic risks to humans, vol 89. IARC, Lyon. https://monographs.iarc.who.int/wpcontent/uploads/2018/06/mono89.pdf
18. IARC (2004) Betel-quid and areca-nut chewing and some areca-nut-derived nitrosamines. In: Monographs on the evaluation of carcinogenic risks to humans, vol 85. IARC, Lyon. https://monographs.iarc.who.int/wpcontent/uploads/2018/06/mono85.pdf
19. IARC (2012) Personal habits and indoor combustions. A review of human carcinogens. In: Monographs on the evaluation of carcinogenic risks to humans, vol 100 E. IARC, Lyon

Risk Factors for Oral Squamous Cell Carcinoma in the Indian Population

2

Suzanne Tanya Nethan, Shalini Gupta, and Saman Warnakulasuriya

Abstract

Oral cancer is caused by a combination of multiple extrinsic and intrinsic risk factors acting over a long period of time. Tobacco is the primary etiological factor for oral cancer; however, this condition may also be associated with various other behavioral risk factors such as areca nut and alcohol consumption, infectious risk factors such as human papillomavirus, and lack of protective factors in the form of inadequate dietary nutrition, to name a few. The risk also varies with gender (higher in males), age (increases with age), and site (lip, oral cavity, or oropharynx). This chapter elaborates upon the various risk factors which contribute to the development of oral cancer.

Keywords

Risk factor · Tobacco · Areca nut · Alcohol · Human papillomavirus · Chronic trauma

---

S. T. Nethan (✉)
School of Preventive Oncology, Patna, Bihar, India
e-mail: suzanne.nethan@gmail.com

S. Gupta
Division of Oral Medicine, Center for Dental Education and Research, All India Institute of Medical Sciences, New Delhi, India
e-mail: shalinigupta@hotmail.com

S. Warnakulasuriya
King's College, London, UK
e-mail: saman.warne@kcl.ac.uk

S. Routray (ed.), *Microbes and Oral Squamous Cell Carcinoma*,
https://doi.org/10.1007/978-981-19-0592-6_2

2.1 Introduction

Oral squamous cell carcinoma (OSCC) is the 17th most common cancer globally [1] and comprises of more than 90% of all the malignant neoplasms of the oral cavity [2]. In India, it is the third most common cancer overall (both genders combined) but the most common cancer among males, with an age-standardized incidence rate of 14.8 per 100,000 [3]. Among cancers contributing to mortality in the country, OSCC ranks third, with an age-standardized mortality rate of 5.4 per annum 100,000 [3].

With respect to gender, males are at a twofold higher risk than females for developing oral cancer [4]. The incidence of oral cancer increases with advancing age. Hence, individuals aged above 60 years are at the highest risk; however, a rise in incidence has also been observed among individuals aged below 45 years probably due to changing risk factors with about 6% cases being reported in cancer registries [5].

A risk factor is defined by the World Health Organisation as "*Any attribute, characteristic, or exposure of an individual which increases the likelihood of developing a disease or injury*" [6]. Oral cancer is caused by multiple extrinsic and intrinsic factors acting over a long period of time [7]. This chapter elaborates upon the various risk factors which contribute to the development of oral cancer. The risk factors for cancers of the lip/mouth and that of the oropharynx also vary as discussed subsequently.

2.1.1 Tobacco

Tobacco is the primary cause of oral cancer, consumed in either the smoked (combustible) or smokeless (non-combustible) form [8]. Globally, almost 1.3 billion tobacco users were reported in 2018, 80% of whom lived in low- and middle-income countries [9]. India is the third largest producer and the second largest consumer of tobacco worldwide [10]. About 28.6% (266.8 million) adults (15 years and above) consume tobacco, with the majority consuming smokeless tobacco (SLT) (199.4 million) followed by smoked tobacco (99.5 million), and a small percentage of tobacco users consuming both the forms/dual users (4%) [10]. The prevalence of current tobacco use observed was 42.4% among men and 14.2% among women. It also showed variation across the states/union territories (UTs), i.e., from as high as 64.5% in Tripura to as low as 9.7% in Goa; the three large states—Uttar Pradesh, West Bengal and Maharashtra—together account for the highest number of tobacco users (more than one third of the total tobacco users in India) [10]. However, tobacco consumption is not restricted to adults but has also shown a significant prevalence among the youth (13–15 years) [11]. Tobacco-related mortality in India is 1.35 million per annum [12] (1.2 million deaths due to smoking while 230,000 deaths due to SLT consumption [13]), oral cancer being one of the main underlying conditions [10].

2.1.2 Smoked Tobacco

Smoked tobacco products include combustible products containing tobacco such as *bidi*, cigarette, hukkah, pipe, cigar, *chillum*, *chutta*, and *dhumti* [10]. In India, *bidi* (slim, hand-rolled, unfiltered cigarettes) [14] is the most popular smoked tobacco product (7.7%) while cigarette is the second most used smoked tobacco product (4%); *hukkah* (water pipe) (0.7%) and other products such as cigars, cigarillos, or cheroots (0.3%) are less popular (Fig. 2.1). Overall, about 10.7% individuals smoke tobacco in India with a greater prevalence among men (19%) than women (2%) [10].

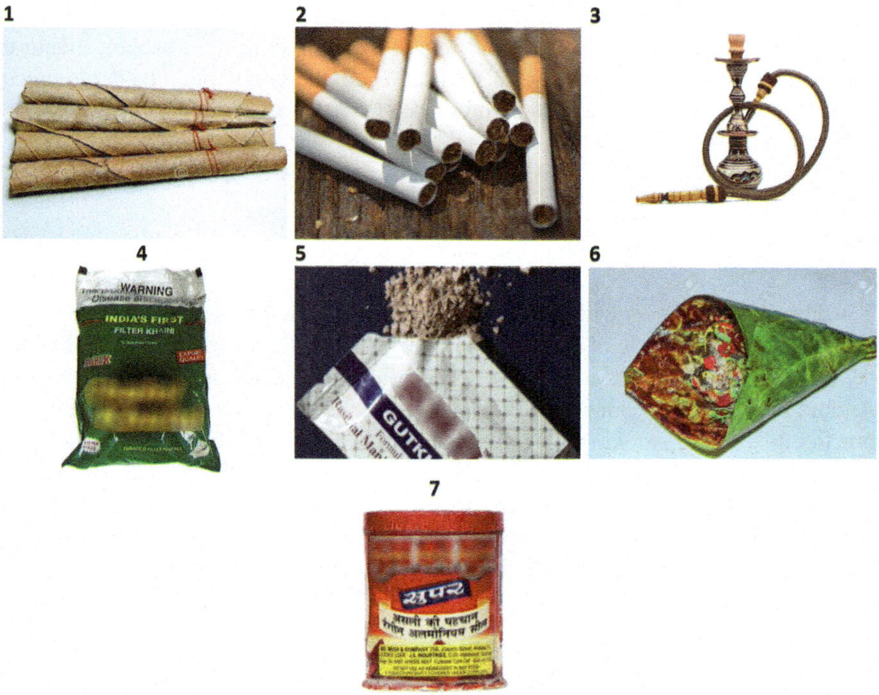

Fig. 2.1 Common tobacco products in India. Smoked tobacco products: (1) *Bidi*, (2) Cigarette, (3) *Hukkah*. Smokeless tobacco products: (4) *Khaini*, (5) *Gutkha*, (6) Betel quid/*Paan*, (7) *Gulmanjan*

2.1.3 E-Cigarettes

An e-cigarette or Electronic Nicotine Delivery System is a battery-operated smoking device containing a solution of nicotine, propylene glycol, glycerin, and flavoring agents [10], which, when heated, produces an aerosol comprising of various human carcinogens such as formaldehyde, acrolein, and acetaldehyde [15, 16]. A rapid increase in the rate of use of e-cigarettes among the youth and adolescents has been noted in many countries where these products were introduced such as in the United States (US) (21% high school and 5% middle school student users in the last 30 days in 2018), Poland (13%), and other European Union countries [17]. In India, among the 3.95% of youth were aware of e-cigarettes, only about 0.54% smoked the same between 2016 and 2017 [10]. Various adverse effects on humans have been documented with the use of e-cigarettes, which include respiratory, cardiovascular, and neurological disorders; cellular, molecular, and immunological toxicity; and negative impact during pregnancy and fetal development [17]. However, owing to their recent introduction and diversity of the available types, the long-term cancer risks of e-cigarette use remain unknown at the population level which warrants further research [17].

2.1.4 Reverse Smoking

Tobacco smoking has a wide regional variation in India, not only in the prevalence [10] but also in the mode of use of the smoked product. Reverse smoking is one such peculiar habit mostly prevalent among the adult fisher community of coastal rural Andhra Pradesh [18], with a much higher prevalence noted among women [19]. The practice of reverse smoking involves placing the lit end of the smoking device—*chutta* (a homemade cigar made by crudely rolling few semi dried tobacco twigs)—inside the mouth, allowing slow inhalation of smoke which is then let out from the unlit end [18]. The commonly affected area is palate, with associated lesions ranging from benign, keratosis potentially malignant, or even malignancy [20].

2.1.5 Second-Hand Smoke (SHS)

Second-hand tobacco smoke is also known as "passive smoking," "involuntary smoking," or "environmental tobacco smoke." SHS comprises the smoke released from the lit end of a smoked tobacco product between puffs (side stream smoke) and that exhaled by the smoker (exhaled mainstream smoke) [21]. SHS is present in virtually all places where smoking takes place, i.e., at home, indoor workplace, or public places such as government buildings, healthcare facilities, restaurants, bars, public transport, and theaters [10]. In India, the highest exposure to nonsmokers to SHS was noted at home (35%), followed by workplace (26.2%), and public places (25.7%) [10].

Biomarker studies have also reported elevated levels of carcinogens in urine; for example tobacco-specific nitrosamines (TSNAs) [22, 23] and polycyclic aromatic hydrocarbons (PAH) [24], in individuals exposed to SHS compared with nonexposed never smokers.

2.1.6 Smokeless Tobacco

A myriad variety of SLT products (Fig. 2.1) are available in the Indian market [25]. Among them, *Khaini* (sun-dried tobacco often consumed with slaked lime) [25] is the most popular SLT product (11.2% users); *Gutkha* (combination of tobacco and areca nut (AN), with additives such as lime and catechu, sweeteners, and other condiments) [25] (6.8% users) and *Paan*/betel quid (BQ) with tobacco (AN combined with tobacco, with various additives such as catechu, alkaline agents, and condiments, all wrapped in betel leaf) [25] (6% users) are also widely consumed [10]. In addition, certain tobacco-based dentifrices, such as *Gul*, *Gudakhu*, *Mishri*, are highly popular as well (3.8% users) (Fig. 2.1). Overall, 21.4% individuals consume SLT with a higher prevalence among males (29.6%) than females (12.8%) [10].

2.2 Risk Estimates

A risk of head and neck cancer (including oral cancer) was reported among smokers (of cigarettes, cigar, pipe—exclusive or joint) by the International Head and Neck Cancer Epidemiology (INHANCE) Consortium through a meta-analysis of 13 case–control studies, i.e., odd's ratio (OR) (95% confidence interval (CI) = 2.42 (1.96–2.99)) (Table 2.1) [26]. Increased risk of head and neck cancers (including oral cancer) with an increase in the frequency (number of cigarettes smoked per day) has been reported, i.e., for 1–10 cigarettes/day, the OR (95% CI) reported was 1.88 (1.72–2.05), and for >10 cigarettes/day, the OR (95% CI) was as high as 4.02 (3.54–4.58) [26]. Another multicentric case–control study in India also reported an OR of 1.91 (95% CI 1.61–2.26) for oral cancer due to cigarette smoking [27] (Table 2.1). Two studies conducted in India also reported a risk as high as OR = 3.1 (95% CI 2.0–5.0) of oral cancer due to *bidi* smoking [28, 29] (Table 2.1). Few studies have reported an association between SHS and oral cavity cancer risk; two multicenter case–control studies conducted in East Asia (eight countries) [30] and Europe (ten countries) [31] showed ORs of 1.83 (95% CI 1.14–2.92) and 5.83 (95% CI 4.50–7.54) (including oropharyngeal cancer), respectively (Table 2.1).

Another recent meta-analysis by Asthana et al. [32] reported a significant risk for oral cancer on SLT consumption in the Southeast Asian Region (the regional analysis included mainly Indian studies) with an OR (95% CI) of 4.44 (3.51–5.61) and a significantly higher risk ($p < 0.001$) among female SLT users (OR [95% CI] = 5.83 [2.93–11.58]) (Table 2.1). Hence, the risk for oral cancer may be higher

Table 2.1 Magnitude of risks of oral cavity cancer due to different lifestyle habits

Habit	Cases/controls	Risk	Variables	Reference/ country of study
Tobacco				
Cigarette smoking	4091 oral cavity cancer cases, 18,664 controls	OR = 2.87 (95% CI = 2.60–3.18)	Adjusted for sex, age, race/ ethnicity, education, frequency of alcohol use (mL/day), duration of cigar smoking (years), and duration pipe smoking (years)	Wyss et al. [26] Multicountry
	1563 oral cavity cancer cases, 3638 controls	OR = 1.91 (95% CI = 1.61–2.26)	Adjusted for age, center, education level, two other habits	Znaor et al. [27] India (multicentric)
Cigar smoking	661 oral cavity cancer cases, 6979 controls	OR = 2.83 (95% CI = 1.91–4.17)	Adjusted for sex, age, race/ ethnicity, education, frequency of alcohol use (mL/day), and duration of pipe smoking (years)	Wyss et al. [26] Multicountry
Pipe smoking	680 oral cavity cancer cases, 7145 controls	OR = 2.51 (95% CI = 1.68–3.75)	Adjusted for sex, age, race/ ethnicity, education, frequency of alcohol use (mL/day), and duration of cigar smoking (years)	
Cigarette, cigar, and pipe smoking (joint)	211 oral cavity cancer cases, 277 controls	OR = 2.42 (95% CI = 1.96–2.99)	Adjusted for sex, age, race, educational level, and frequency of alcohol use	
Bidi smoking	6270 oral cavity cancer cases, more than 4438 controls	OR = 3.1 (95% CI = 2.0–5.0)	None	Rahman et al. [28] Indian subcontinent (mainly India)

(continued)

Table 2.1 (continued)

Habit	Cases/controls	Risk	Variables	Reference/ country of study
	282 oral cavity cancer cases, 1410 controls	OR = 1.9 (95% CI = 1.1–3.2)	Adjusted for education, religion, chewing, and alcohol drinking habits	Muwonge et al. [29] India
Involuntary smoking/ secondhand smoke	303 oral cavity cancer cases, 459 controls	OR = 1.83 (95% CI = 1.14–2.92)	Adjusted for center, age, sex, race/ethnicity, education, and alcohol drinking frequency	Hashibe et al. [30] East Asia (multicentric)
	974 oral cavity and oropharynx cancer cases, 2168 controls	OR = 5.83 (95% CI = 4.50–7.54)	Adjusted for age, sex, education, center, and alcohol drinking frequency (continuous) and duration (continuous)	Lee et al.[a] [31] Europe (multicentric)
Smokeless tobacco (Betel Quid (BQ), *Pan* Snuff, *Gutkha*, *Supari*, *Mishri*, betel leaf, *Khaini*, areca nut, *Zarda*, lime, snuff, *Mainpuri*, oral snus, fermented)	24 case–control studies: 6869 oral cavity cancer cases, 32,764 controls.	OR = 4.44 (95% CI = 3.51–5.61)	Adjusted at least for smoking.	Asthana et al. [32]
	4 cohort studies: 224,340 participants		Others (one or more): alcohol, age, body mass index, and geographical region	India (majority), Indonesia, and other Central and Southeast Asian countries
Areca nut-based products				
BQ without tobacco[a]	30 case–control studies: 8012 oral cavities and/or oropharyngeal cases, 10,450 controls.	Indian subcontinent: metaRR = 3.22 (95% CI = 2.11–4.92)	Adjusted for smoking and alcohol	Guha et al.[a] [33]
	1 cohort study: 57,518 oral cavity cancer/ precancerous cases			Indian subcontinent (mainly India) and Taiwan

(continued)

Table 2.1 (continued)

Habit	Cases/controls	Risk	Variables	Reference/ country of study
BQ with tobacco[a]	12 case–control studies: 13,608 oral cavities and/or oropharyngeal cases, 444,824 controls.	metaRR = 7.03 (95% CI = 4.68–10.56)	Adjusted for smoking, and alcohol	Guha et al.[a] [33]
	3 cohort studies: 110,583 oral cavities and/or oropharyngeal cancer and/or precancer cases			Indian subcontinent (mainly India)
Alcohol	8 case–control studies: 3955 oral cavity cancer cases, 8614 controls.	Relative risk (cohort studies) = 3–9	Adjusted for smoking	Goldstein et al. [34]
	12 cohort studies: 578 oral cavity cancer cases	OR (case–control studies) = 2–14		Systematic review
	17,085 oral cavities and/or pharyngeal cancer cases, 2,934,880 controls	RR = 1.21 (95% CI = 1.10–1.33) for less than one drink (10 grams ethanol) per day	Adjusted for tobacco, social class, selected dietary factors, oral hygiene, and other recognized factors, where available	Tramacere[a] et al. [35]
		RR = 5.24 (95% CI = 4.36–6.30) for at least 4 drinks per day		Multicountry
	4471 oral cavity cancer cases, 6768 controls	OR = 1.92 (95% CI = 1.54–3.96)	Not specified	Singhavi et al. [36]
				India
	1563 oral cavity cancer cases, 3638 controls	OR = 1.98 (95% CI = 1.68–2.33)	Adjusted for age, center, education level, two other habits	Znaor et al. [27]
				India (multicentric)

OR odds ratio, *CI* confidence interval, *RR* risk ratio
[a] Some studies include oropharyngeal cancers

among SLT users than tobacco smokers. With regard to the different types of SLT products, the relative risk for oral cancer was higher for chewable (OR [95% CI] = 4.37 [3.27–5.83], $p < 0.001$) than nonchewable SLT products (OR [95% CI] = 1.56 [1.04–2.36], $p = 0.03$), i.e., highest for *Gutkha* (OR [95% CI] = 8.6

[3.59–20.93]) followed by Paan with tobacco (OR [95% CI] = 7.18 [5.48–9.41]) [32].

2.3 Carcinogens in Tobacco Products

Tobacco smoke comprises of about 5300 toxic chemical compounds among which 70 are proven carcinogens in either laboratory animals or humans, most important being PAH, TSNAs, aromatic amines, aldehydes, and certain volatile organics. The level of TSNAs has been found to be similar in commercial Indian cigarettes and *bidis* [8].

SLT products comprise about 4000 similar toxic chemical constituents (including carcinogens) among which TSNAs are considered the most potent carcinogens, especially 4-(methylnitrosamino)-1-(3-pyridyl)-1-butanone (NNK) and *N*-′-nitrosonornicotine (NNN). A wide variation in TSNA levels has been noted in the different Indian SLT products, i.e., from 264 to 23,900 ng/g in *Gutkha* up to 21,600–23,500 ng/g in *Khaini* [25].

2.3.1 TSNA production

Nitrate (NO_3^-) is absorbed and metabolized by tobacco plants from soil and fertilizers, as they grow. At the plant's harvesting, it subsequently undergoes microbial conversion to nitrite (NO_2^-), which begins once the tobacco leaves are subjected to curing (drying). The nitrite thus produced reacts with tobacco alkaloids and other secondary/tertiary amines (nitrosation) to generate TSNAs [25].

2.3.2 Areca Nut

Areca nut (AN) is the seed of the fruit of the Areca catechu palm, native to Southeast Asia and East Africa [37]. It is the fourth most used addictive substance with almost 600 million users globally [38].

Areca nut has been designated as a Class I carcinogen by the International Agency for Research on Cancer (IARC). It causes oral submucous fibrosis, an OPMD, in humans, in addition to showing sufficient evidence of carcinogenicity in mechanistic studies and in experimental animals. In addition, it is also an integral component of BQ which has been designated as a Class I carcinogen as well, owing mainly to the presence of AN and SLT [8, 39].

AN may be consumed ripe or unripe; roasted, dried, or baked; crushed, sliced, or whole; and with or without tobacco [40]. The common AN product consumed in India are *Paan*/BQ (without tobacco)—8.7% users, whole AN—8% users, and *Pan Masala* (mixture of AN, slaked lime, and other flavoring agents)—4.8% users. Like tobacco, even most AN products are more commonly used by males; however, the consumption of *Paan* (without tobacco) is slightly higher among females (9%) than males (8.4%). A large state-wide is noted in the prevalence of AN products with the North-Eastern states showing the highest consumption [10]. Children usually begin deleterious chewing habits with AN product, especially *Supari* (sweetened and flavored AN) with many switching to tobacco use subsequently. Thus, AN use may also act as a gateway for tobacco use [41].

A meta-analysis of 13 case–control studies conducted in the Indian subcontinent showed *Paan*/BQ without tobacco as an independent, strong risk factor for oral and/or oropharyngeal cancer, i.e., meta-risk ratio (mRR) (95% CI) = 3.22 (2.11–4.92) (Table 2.1) [33]. Subgroup analysis of seven of these studies showed a higher risk for female chewers—mRR (95% CI) = 3.02 (1.79–5.08); the risk of oral cancer was also directly proportional to the amount consumed daily and the duration of chewing BQ [33].

2.3.2.1 Carcinogens in AN products

Areca-specific alkaloids, arecoline, guvacine, guvacoline, and arecaidine have been implicated in the carcinogenicity of AN through areca-derived nitrosamines (*N*-nitrosoguvacine, *N*-nitrosoguvacoline, and 3-methylnitrosamino-propionitrile), formed due to the reaction with nitrite (because of conversion of dietary nitrate) present in saliva [42]. However, recently, arecoline was classified as being "possibly" carcinogenic in humans (Group 2B) based on strong mechanistic evidence; no evidence regarding cancer in humans was available for arecoline [43]. During the processing of AN or the preparation of products like BQ, *Gutkha* or *Pan Masala*, other harmful chemicals may be introduced, examples: PAH formation during AN roasting, alkaline agents like calcium hydroxide to enhance absorption of the product, nicotine and TSNAs due to the added tobacco, and other toxic chemicals present in the flavoring and coloring agents [42].

2.3.3 Alcohol

Alcohol is the second most widely used psychoactive substance [38] with an addictive potential. In the current age, alcoholic beverages frequently accompany socializing [44], resulting in increasing consumption levels. Globally, in 2016, alcohol per capita (APC) consumption was estimated to be 6.4 L (pure alcohol) per person aged 15 years or above, with the highest consumption average noted in Europe (9.8 L) in the Republic of Moldova (15.2 L) [45]. Regarding heavy episodic drinking or "binge drinking" (60 g or above of pure alcohol on at least one single occasion at least once per month), the APC was estimated to be 6.4 L among the total population and 15.1 L among drinkers, with the highest prevalence in the European region (9.8 L) [46].

India is the third largest market for alcohol worldwide; the reported market size was US$ 35 billion in 2017, which was estimated to increase up to US$ 52 billion by 2020. The states of Uttar Pradesh, Maharashtra, Tamil Nadu, and Karnataka contribute to 62% of the total alcohol production in the country [47]. A notable increase of 38% was recorded in the per-capita consumption of alcohol between 2010 and

2017, i.e., from 4.3 to 5.9 L. The prevalence of current alcohol use was reported as 14.6% which was much higher among males (27.3%) than females (1.6%). A wide variation across the states/UTs was also noted; some states reported high alcohol consumption even among females, such as Arunachal Pradesh (26.3%), Sikkim (23.0%), Telangana (8.7%), Chhattisgarh (5.0%), and other North Eastern states [48, 49]. With respect to the most common types of alcoholic beverages consumed in India, hard liquor containing high alcohol content is the main preference with 60% individuals consuming either country liquor or spirits (Indian-made foreign liquor); 21% and 4% users reported consuming beer and wine, respectively [48].

Alcohol (primarily ethanol) has also been designated as a class 1 carcinogen by the IARC, causative for oral cancer [50]. Goldstein et al., in their global meta-analysis, showed alcohol consumption to have an independent effect on the risk of cancer of the oral cavity with relative risk ranging from 3 to 9 in 12 cohort studies and OR ranging from 2 to 14 in eight case–control studies (Table 2.1); a strong dose–response relationship was also observed in all the studies [34]. The association was seen across different geographic locations and populations, which further supports this evidence [34]. In addition, the dominant alcohol types in specific regions were associated with the greatest risk [34]. In a meta-analysis of two cohort and 43 case–control studies, the pooled RR (95% CI) ranged from 1.21 (1.10–1.33) for less than one drink (10 g ethanol) per day up to 5.24 (4.36–6.30) for heavy alcohol use (at least four drinks per day) (Table 2.1) [35, 51]. According to a recent meta-analysis of seven Indian studies by Singhavi et al., the oral cancer risk is double among alcohol consumers, i.e., pooled OR (95% CI) = 1.92 (1.54–3.96) than non-drinkers (Table 2.1) [36]. A multicentric case–control study in India also reported a risk for oral cavity cancer due to alcohol consumption, i.e., OR = 1.98 (95% CI 1.68–2.33) (Table 2.1) [27].

2.3.3.1 Mechanism of Action

The following mechanisms have been proposed by which alcohol/ethanol contributes to oral carcinogenesis [52]:

- Alcohol may influence the proliferative cells by both intracellular (endocytosis) and intercellular (permeability) pathways by increasing mucosal permeability, thus facilitating the penetration of carcinogens into the tissues [53, 54].
- Alters the mucosal morphology with a resultant reduction in the epithelial thickness.
- Through its metabolite—acetaldehyde (produced by the oral bacteria and oral epithelial cells) [55], which is a Class I carcinogen as well and may also potentiate the carcinogenic effects of other agents. Acetaldehyde protein adducts have been demonstrated in oral cancer tissues in heavy alcohol drinkers [56]. Acetaldehyde may also disrupt salivary gland function, reducing the salivary flow, thus impacting the clearance of locally acting carcinogens.

2.3.4 Tobacco–Alcohol Synergism

A synergistic effect of alcohol when consumed with tobacco, on the occurrence of OSCC, has also been reported in the literature (OR [95% CI] = 5.37 [3.54–8.14]). The greatest odds for OSCC occurrence were observed with the synergistic consumption of alcohol, smoked tobacco, and SLT (OR [95% CI] = 16.17 [7.97–32.79]), followed by the combination of alcohol and SLT consumption (OR [95% CI] = 7.78 [2.86–21.14]), and alcohol and smoked tobacco (OR [95% CI] = 4.74 [3.51–6.40]) [57].

2.4 Socioeconomic Factors

Socioeconomic status is an independent risk factor and acts synergistically with major risk factors like tobacco, areca nut, and alcohol consumption in the development of oral cancer. Low socioeconomic status is measured by occupation, income, or education. Lack of access to healthcare facilities, poor hygiene, poor nutrition, unfavorable work environment and poor living conditions in general also contribute to the overall risk of oral cancer [58, 59]. Lack of awareness about oral cancer screening was also found to be related to education levels race, income, immigration status, and health insurance status [60]. Oral cancer patients with low socioeconomic status are also diagnosed at later stage of disease due to delay in seeking treatment and have poorer survival rates [61, 62]. Oral cancer patients with low socioeconomic status from disadvantaged neighborhoods have a 1.46- to 1.64-fold higher risk of death as compared to oral cancer patients with high socioeconomic status from advantaged neighborhoods [63]. Improving health literacy on risk factors of oral cancer and its early detection through screening programs should be directed toward people with low socioeconomic status.

2.5 Microbes

2.5.1 Human Papillomavirus

Human Papilloma vIrus (HPV) is a nonenveloped double-stranded DNA epitheliotropic virus that is responsible for many benign and malignant mucosal and cutaneous neoplasms. More than 200 genotypes of HPV have been found and are usually grouped into high-risk and low-risk HPV types. Low-risk HPV types include types 6, 11, 42, 43, and 44. High-risk oncogenic HPV types include types 16, 18, 31, 33, 34, 35, 39, 45, 51, 52, 56, 58, 59, 66, 68, and 70. Some HPV types are often found in squamous intraepithelial lesions and are referred as intermediate risk. HPV 16 and 18 are usually associated with head and neck cancers. In oral carcinogenesis, HPV usually infects the basal epithelial cells through micro-abrasions or surface wounds and gets integrated in the host genome where increased expression of viral oncoproteins E6 and E7 causes deactivation of tumor suppressor proteins p53 and retinoblastoma pRb. This leads to dysregulation of cell cycle and immortalization of keratinocytes. Most of HPV infections are transient and are cleared in 12–24 months, but persistent HPV infection could lead to the development of carcinomas especially in immunocompromised individuals with HIV infection [64]. It is usually linked to sexual transmission but horizontal, mouth-to-mouth, and vertical transmissions are also possible [65]. Nearly 70% of oropharyngeal cancers are HPV positive, the most common sites being the base of tongue and tonsils. In 2011, the International Agency of Research of Cancer (IARC) has acknowledged the epidemiological evidence for the role of HPV 16 and possibly HPV18 in the etiology of oral cancer [66]. The prevalence of HPV in oral cancer globally is 24.4% while in Indian patients it is 36.6% [67]. Tongue (50%), palate (42%), gingiva (40%), and floor of mouth (26.8%) are common sites of oral HPV infection [67]. Deep periodontal pockets have been found to provide a niche for HPV, EBV, and HSV viruses. The periodontium may serve as reservoir for latent HPV infection as E6/E7 mRNA has been found there [68].

HPV-positive head and neck cancers are associated with advanced clinical stage, lower tumor volume and aggressive nodal involvement as compared to HPV-negative cancers. Histologically, HPV-positive head and neck cancers have moderate/poor tumor differentiation and non-keratinizing or basaloid pathology. In Indian patients, most of the HPV-positive head and neck cancers were found to be T3/T4; stage of the disease with high nodal involvement and well-differentiated tumors in contrast to western literature. Overexpression of $p16^{INK4A}$ determined by immunohistochemistry (p16 positive) is highly correlated to HPV infection in head and neck cancers and considered as a surrogate marker for HPV positivity. No survival difference has been found between p16-positive and p16-negative head and neck cancers in Indian patients. The difference in clinical presentation and survival of HPV-positive head and neck cancers in Indian patients could be attributed to tobacco usage in these patients [69]. The overall pooled prevalence of HPV in oral potentially malignant disorders (OPMD) has been found to be 22.5% and comparable across oral leukoplakia, oral lichen planus, oral submucous fibrosis, and proliferative verrucous leukoplakia, independent of dysplasia. HPV 16 was the predominant genotype in HPV-positive OPMD [70]. There is however great heterogeneity in the studies related to HPV prevalence in oral cancer and OPMDs from different populations due to differences in geographical regions, ethnicity, HPV detection, and sampling methods. Prophylactic HPV vaccination has been recommended in adolescent girls and boys in several countries as part of national immunization program to prevent HPV-related benign and malignant lesions, and the beneficial effects will be revealed in future generations.

2.5.2 Epstein Barr Virus (EBV)

EBV virus is a ubiquitous double-stranded DNA, gamma-herpes virus that exhibits a dual tropism for B lymphocytes and epithelial cells. It has been found to be associated with several lymphomas and carcinomas at various anatomical sites.

The lymphoid-rich regions like the tonsils and adenoids of Waldeyer's ring in the oral cavity are the primary sites for residence, transmission, and persistence of EBV, from where it is periodically shed in saliva throughout the lifetime of the infected individual. The epithelial cells help in the amplification of the virus while the B lymphocytes are the reservoirs that support the latent phase of the viral cycle for persistent EBV infection while evading immune detection. The oncogenic potential of the EBV virus is believed to be due to the proliferative potential of the immortalized EBV-infected B lymphocytes and epigenetic reprogramming of infected tumor cells that helps in tumor progression. EBV latent transcripts (latent membrane protein 1, EBV nuclear antigen 1 and 2, and EBV-encoded small RNAs) also have an important role in this process [71–73]. The risk of oncogenesis is increased in immunosuppression caused due to HIV infection or organ transplantation. Diffuse large-cell B lymphomas and Burkitt's lymphoma are the most common EBV-associated non-Hodgkin lymphomas that could manifest in the oral cavity. EBV has also been found to be associated with OSCC, and the evidence for the strength of the association has increased in recent years [74]. The prevalence of EBV in oral cancer has been found to be 45.37% while the pooled odds ratio of EBV infection and risk of OSCC with 95% CI was found to be 5.03 (1.80–14.01) [72, 74]. EBV-related OSCC is poorly differentiated and shows enhanced invasiveness. However, there was a great heterogeneity in the studies reported due to geographical, ethnic, EBV detection, and sampling method variations. Larger sample size and high-quality studies are required to confirm the association. EBV has been found as a coinfection with HPV in 15–20% of oropharyngeal carcinoma, 25% of tonsillar, and 70% of tongue cancers and is most likely in cells harboring integrated HPV. The influence of EBV and HPV on one another's viral life cycles and oral carcinogenesis needs further investigation [73–75].

2.5.3 Candida

Candida albicans is a commensal of the oral cavity which becomes pathogenic due to changes and weakening of local and systemic host immune responses [76]. The relation between cancer and candidal infections was related to the immunosuppression resulting from chemotherapy and radiotherapy for cancer and invasive infections leading to hematogenous dissemination causing life-threatening candidemia. The complications with candidal infections in oral cancer were associated with higher mortality, prolonged hospitalization, and increased healthcare costs. However, recent findings demonstrate that *Candida albicans* can also promote oral carcinogenesis by several mechanisms that can cause induction, progression, and metastasis of tumors. *Candida* has been isolated from various OPMDs and oral cancer and found to be associated with epithelial dysplasia. *Candida* infection in chronic hyperplastic candidosis has been found to be associated with malignant transformation [76]. A recent systematic review of 16 case–control and cohort studies has found candida infection to be present in 6.8–100% of oral leukoplakia while three studies reported a malignant transformation rate of 2.5%, 6.5%, and

28.7% in oral leukoplakia infected with *Candida* [77]. *Candida albicans* can produce nitrosamines that activate specific proto-oncogenes that can cause development of dysplasia in oral epithelium or malignant lesions. *Candida albicans* can metabolize ethanol with the enzyme alcohol dehydrogenase (ADH) and produce acetaldehyde. Acetaldehyde can produce DNA and protein adducts, aberrant molecules with structural and functional alterations causing point mutations and chromosomal aberrations. Acetaldehyde also causes increase in reactive oxygen species (ROS) by binding to glutathione and causes increase in DNA and mitochondrial damage leading to increase in cell apoptosis and tumor progression. *Candida albicans* can produce high levels of acetaldehyde in unfavorable conditions like low oxygen concentrations such as poor oral hygiene which is common in tobacco and alcohol users. *Candida albicans* can induce an inflammatory response from endothelial cells and oral epithelial cells by cytokine production to promote metastasis and cause increase in tumor cell adhesion by alterations in adhesion molecule expression. It has also been proposed that Th-17 cells, the dominant response against *Candida albicans*, produce IL-17 which not only plays a role in antifungal immunity but also in carcinogenesis. The complement receptor-3-related protein (CR3-RP) of *Candida albicans* and the complement receptor 3 (CR3) required for adhesion of leukocytes to the endothelium for their subsequent extravasation share antigenic and structural similarities. These similarities, also called molecular mimicry, is another mechanism that has been postulated for carcinogenesis. The anti-candida and anti-tumor defensive response of the host can be affected by cross reactivity of antibodies against CR3-RP of *Candida albicans* with CR-3 of leukocytes [78]. The interactions of *Candida albicans* with other microbes in the oral cavity not only increases the ability to form biofilms but also promotes carcinogenesis by exacerbating the inflammatory response and affecting attachment to extracellular matrix molecules (ECM) and epithelial–mesenchymal transition (EMT) of malignant cells [79]. Candidalysin a fungal toxin peptide, encoded by *ECE1* gene, secreted by invasive *Candida albicans* hyphae, promotes tissue invasion and epithelial damage leading to activation of several immune response pathways associated with cell proliferation, survival, angiogenesis, differentiation, and motility leading to carcinogenesis [80]. Further studies are required to delineate the exact role of *Candida* in oral cancer and malignant transformation of OPMDs. However, identification and elimination of predisposing factors for *Candida* infection in OPMDs and oral cancer can be undertaken as a preventive measure.

2.5.4 Herpes Simplex Virus (HSV)

HSV1 is an enveloped double-stranded DNA virus that is primarily associated with oral and ocular infections. HSV2 is mainly associated with genital infections. HSV1 antibody titers are found to be higher in oral cancer patients especially IgA and IgM. High levels of IgM antibodies were associated with shorter survival. History of HSV seropositivity and cigarette smoking have been found to increase the risk of oral cancer. Some preliminary reports have also found the presence of HSV viral DNA and RNA in oral cancer. The ability of HSV to transform cells during infection is caused by the induction of cellular proteins. Cell transformation can also be related to the ability of HSV to stimulate other viruses like HPV or exposure to chemical carcinogens. Chromosomal damage at sites on chromosome 1q and on chromosome 3, 9, 16 might be another mechanism of cell transformation by HSV. A possible hit and run mechanism for HSV in oral carcinogenesis has also been postulated where HSV2 may act as promoter or initiator in early neoplastic changes in well-differentiated OSCC that becomes denatured and less evident in less differentiated tumors [81]. A synergistic role between HSV1 and HPV in early oral carcinogenesis has also been proposed. There is an increase in incidence of OSCC in young, nonsmoking, nondrinking men and women in western countries. The epidemiological, clinical, and molecular profile of these tumors suggest a viral etiology and has been hypothesized as being caused by changes in sexual behaviors and increased incidence of oral herpesvirus infection especially HSV2 [82]. HSV1 has also been found to be associated with OPMDs like OSMF and OLP but not significantly, and their role in malignant transformation is not known [83–85]. Since HSV1 is strongly cytolytic, its replication restricted to tumor cells can help in the treatment of oral cancer [86]. Genetically engineered HSV1 genes have been used as oncolytic virus against head and neck cancer including oral cancer [87].

2.5.5 Periodontal Pathogens

Tobacco smoking, smokeless tobacco, areca nut chewing habits, and alcohol consumption are not only associated with increased risk of oral cancer but also found to be associated with poor periodontal health [88–91]. Poor oral hygiene, high plaque indices, gingival inflammation, gingival recession, periodontal pockets, furcation involvement, and mobility of teeth are commonly seen in users of these products. The highly addictive nicotine in tobacco has been found to be cytotoxic to human gingival, periodontal, and oral epithelial cells [92]. Tobacco also aggravates periodontitis by promoting invasion of periodontal pathogens and inflammatory reaction by the inhibition of immune defense barriers and causing loss of alveolar bone [93].

Periodontal diseases, tooth loss, and alveolar bone loss have been considered as risk factors for oral cancer [94]. The oral biofilms contain many pathogenic and commensal microflora that cause periodontitis and are also implicated in oral carcinogenesis. *Aggregatibacter actinomycetemcomitans*, *Porphyromonas gingivalis*, and *Tannerella forsythia* are most common bacteria in periodontitis. Actinomycetemcomitans is associated with gastric precancers while *Fusobacterium* species are associated with esophageal, colorectal, and pancreatic cancers. The periodontal pathogens most isolated from oral cancers are *Fusobacterium nucleatum* and *Porphyromonas gingivalis*. Cervical lymph nodes in patients with oral cancer have been found to harbor oral bacteria like streptococci (*S. intermedius*, *S. constellatus*, *S. oralis*, *S. mitis*, *S. sanguis*, *S. ssalivarius*) [8] *H. pylori* has been found to be associated with periodontitis and nearly 40% of head and neck cancers.

The association is greater in laryngeal and pharyngeal cancers as compared to oral cancer [95].

P. gingivalis activates some immunologic and inflammatory reactions, invades and penetrates the oral epithelium, and triggers NFkB and MAPK pathways. It promotes the secretion of IL-6 and targets CD274 and programmed cell death 1 ligand 2 (PDCD1LG2) to activate tumorigenic transcription factors like STAT1. It can induce MMP-9 that can degrade the tumor microenvironment to cause invasion and metastasis of neoplastic cells. It can also induce epithelial–mesenchymal transition (EMT) of malignant epithelial cells, neoplastic proliferation and tumor invasion [96]. The chronic inflammation set up by the periodontal pathogens activates various inflammatory cells, chemokines and cytokines that contribute to an inflammatory microenvironment in which carcinogenesis can occur [68, 96, 97]. The oral microbiota can also produce reactive oxygen and nitrogen species and nitrosoamines inducing genetic damage of oral epithelial cells. They can also cause dysfunction of epithelial barriers leading to alterations in microenvironment and predisposition toward the development of oral cancer. Dysbiosis of oral microflora may cause several epigenetic changes by producing different metabolites affecting the host cell metabolism and alterations in gene and miRNA expressions (Fig. 2.2) [98]. Certain bacterial strains (*Lactobacillus rhamnosus GG* (LGG), *Lactobacillus plantarum*) may have a protective role against oral carcinogenesis (Probiotics) by reducing the mutagenic effects of harmful substances, thus maintaining the oral mucosa homeostasis and may serve as novel preventive and therapeutic strategies in the future [98]. Maintenance of oral hygiene and early treatment and control of periodontal diseases should be a part of primary prevention of oral cancer.

2.6 Chronic Trauma

Chronic trauma and mechanical irritation have been proposed as an etiological factor for oral cancer in retrospective case–control and descriptive studies though the evidence at present is limited and controversial [99]. The mechanisms for carcinogenesis due to chronic trauma may be related to DNA damage of cells and inflammation leading to release of cytokines, prostaglandins, and tumor necrosis factor, leading to oxidative stress, genetic and epigenetic changes, preventing apoptosis and stimulation of angiogenesis. The chronic trauma has been attributed to poor dentition, missing teeth, ill-fitting dentures, and dental implants. Chronic parafunctional habits like lip and cheek biting or sucking and tongue thrusting could also be responsible for chronic irritation [100]. The most common sites for cancers arising due to chronic dental trauma are the lateral border of tongue and alveolar arch. Denture use and more specifically ill-fitting dentures can cause oral cancers of the alveolar arch especially in the mandibular arch. The duration of denture use does not seem to have any relation to the development of cancer. Dentures may also act as reservoirs for *Candida* and HPV and may contribute to carcinogenesis. Monomers leached from the denture base have also been implicated as they can cause irritation,

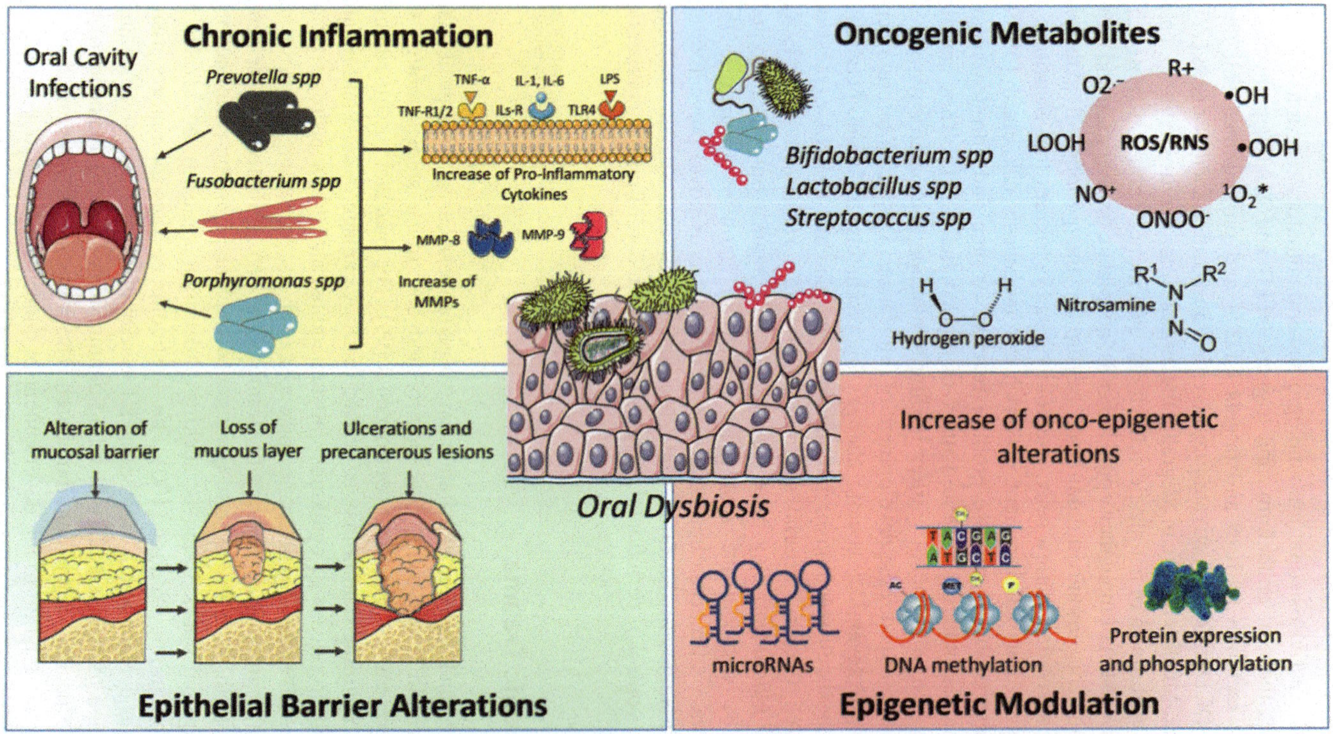

Fig. 2.2 Oral microbiota dysbiosis is associated with oral cancer development through different mechanisms. Oral infections and dysbiosis are responsible for the instauration of a pro-inflammatory microenvironment of which inflammatory cytokines and matrix metalloproteinases favor the development and progression of tumors. Furthermore, the bacteria in the oral cavity produces oxygen and nitrogen reactive species, as well as oncogenic metabolites (e.g., nitrosamines) to induce genetic damage to cells composing the oral mucosa. Another mechanism of neoplastic transformation mediated by oral dysbiosis is the

inflammation, and allergic reactions. Sharp, broken, or defective teeth have also been found in association with oral cancer but not significantly. Some cases of oral cancers arising in close association with dental implants have been reported in patients with previous history of cancer. The release of corrosive products, metallic ions, and migration of malignant cells through the sulcus around implants have been postulated as possible mechanisms of carcinogenesis. Trauma-associated oral cancers usually present with early nodal disease and are more likely to occur in females who do not have any addictive habits of tobacco and alcohol [101]. No association between chronic trauma and OPMD has been reported in a retrospective study; however, oral carcinoma arising in OSMF usually occurs in the buccal mucosa and retromolar trigone and is attributed to chronic trauma due to continuous chewing [102]. Removal and treatment of chronic irritational factors (dental, prosthetic, and functional) in susceptible individuals has been recommended to reduce the risk of oral cancer [103].

2.7 Others

2.7.1 Mouthwash Use

Mouthwashes are routinely used to mask halitosis and for plaque control. Many mouthwashes contain alcohol in the concentrations of 5–27%, which acts as a solubilizer, stabilizer, preservative, to enhance the antiplaque efficacy and as a flavoring agent [104]. Owing to their alcohol/ethanol content, some mouthwashes have been suggested as having a role in oral cancer development. The proposed mechanism of action is like that of alcoholic beverages as mentioned in the previous section [52].

A meta-analysis of 16 case–control studies on mouthwash (including that containing >25% alcohol) and oral cancer showed neither a statistically significant association between its use and risk of oral cancer nor a significant trend in the risk with increasing daily use; no association was seen even between the use of alcohol-containing mouthwash and oral cancer risk [104]. Also, such a risk may be solely seen in smokers [105]. Hence, there is still insufficient evidence regarding the role of mouthwash use as an independent risk factor for cancer of the oral cavity.

Fig. 2.2 (continued) alteration of the epithelial barriers predisposing the individuals for the development of chronic pre-cancerous lesions. Finally, oral dysbiosis is responsible for several epigenetic alterations predisposing the development of tumors (e.g., alteration of onco-miR or DNA methylation phenomena). (Reproduced with kind permission from Oncology Letters (2019), Association of oral dysbiosis with oral cancer development (Review) Giusy Rita Maria La Rosa et al., DOI: 10.3892/ol.2020.11441 Spandidos Publications UK Ltd, 5-6 King Street, Cloisters, Clifton Walk, London W6 0GY, United Kingdom)

2.7.2 Hot Drinks

Consumption of very hot drinks, a practice in some cultures, probably also increases the risk of cancer including oral cancer [106]. Drinking very hot beverages at above 65 °C was classified as Group 2A carcinogens (probably carcinogenic to humans) by the IARC in 2016. The possible mechanism involved may be the resultant cell injury and the sequelae that might lead to cancer. A hospital-based case–control study in Maharashtra (India) showed an increased risk of oral cancer on drinking very hot tea (OR [95%CI] = 3.19 [1.99–5.11]) [107].

Maté, an infusion prepared from dried leaves of *Ilex paraguariensis*, is a traditional drink of South America which is consumed very hot (>65 °C) [108] and thus has been extensively discussed as a possible carcinogen to humans by the IARC [109]. A meta-analysis of four case–control studies on *maté* drinking and oral cancer from Latin America reported a significant OR (95% CI) = 2.11 (1.39–3.19) and a population attributable risk of 16%; however, no significant difference was seen in associated risk between *maté* being drunk as a hot or cold beverage [110]. An increased odds was observed regarding *maté* consumption and overall occurrence of upper aero-digestive tract cancers (OR [95% CI] = 2.24 [1.74–2.87]) in a meta-analysis of 15 case–control studies [111].

2.7.3 Occupational Exposure

The IARC Monographs identify occupational exposure among the environmental factors that can increase the risk of cancer in humans [112, 113]. Occupational exposure, as defined by the Occupational Safety and Health Administration, US, refers to "A reasonably anticipated skin, eye, mucous membrane, or parenteral contact with blood or other potentially infectious materials that may result from the performance of an employee's duties" [114].

Although occupation is not traditionally considered a common risk factor for oral cancer, a recent literature review by Awan et al. [115] demonstrated the role of occupational exposure to certain substances for an increased risk for oral cancer. Asbestos (hazards ratio = 2.47 [1.89–3.21]) [116] and wood particles (OR = 5.5 [1.2–25]) [117] were associated with a higher risk of oral cancer while the observed association between oral cancer with coal particles [118], and formaldehyde [119, 120] was not significant.

Among white-skinned populations, outdoor occupations (e.g., farming and fishing) due to chronic exposure to UV light have been linked with actinic keratosis of lower lip—a potentially malignant disorder (OPMD) that may transform to cancer.

2.7.4 Genetic Predisposition

Inherited or acquired genetic factors may also have an etiologic role in the development of oral cancer. Nearly 5% of the oral cancers are caused by inherited cancer syndromes which have mutations in oncogenes, tumor suppressor genes, and genes involved in angiogenesis. Knudson's two-hit hypothesis helps to explain cancer inheritance where the first hit is the inherited mutation while the second hit occurs later in life. These inherited cancer syndromes have a lifetime high risk of developing cancers which have an early onset and development of OPMDs and multiple independent primary tumors. Dyskeratosis congenita, severe generalized recessive dystrophic epidermolysis bullosa, xeroderma pigmentosum, and Fanconi's anemia are some inherited syndromes that increase the risk for oral cancer [121–123]. Familial aggregation has also been reported in oral cancer with possibly an autosomal dominant mode of inheritance [124].

Genetic variations due to processes like segregation of chromosomes, genomic copy number, loss of heterozygosity, regulations of cell-cycle checkpoints, DNA damage repairs, telomere stabilities, and defects in notch signaling pathways can lead to oral carcinogenesis. Twenty-eight genetic loci associated with oral cancer have been identified by candidate gene studies, genome-wide association studies, and next-generation sequencing approaches [125]. Chromosomal aberrations linked to important target genes like TP53 in 17p13, RB1 in 13q14, CDKN2A in 9p21, loss of heterozygosity in 3p, 9p, 13q, and 17p have been implicated. Dysregulated oncogenes like ErbB, EGFR, CCND1, VEGF, MMP, TIMP, tumor suppressor genes like p53, and pRb are commonly associated with oral cancer [126]. Development of oral cancer is seen in only 5–10% of individuals with the traditional high-risk factors; hence, inherent individual genomic variations such as mutations, deletions amplifications, and single-nucleotide polymorphisms also constitute important risk factors for oral carcinogenesis. A meta-analysis of various studies has identified nine single-nucleotide polymorphisms (SNPs) associated with critical functions deregulated in oral cancers including cell proliferation, inflammation, immune function, DNA repair, transcription, and xenobiotic metabolism [127]. SNPs like rs1800471 (TGF-β), rs1800870 (IL-10), rs20417 (COX-2), rs11549467 (HIF), and GSTM1 null, with the confidence interval range of 1.2–4.2, have been found to be associated with an increased risk of oral cancer [127]. SNPs like COX-2 -1195G/A, GSTM1 null genotype, hMLH1 -93 A>G, IL4-589 CC, -33 CC genotype, and *C allele, 869 CC genotype, and C allele of TGFβ1 have also been found in association with oral cancer in Indian patients [127]. However, a panel of SNPs may be required to effectively work as prognostic markers for oral cancer.

Single case–control studies have found SNPs related to HIF-1α, VEGF-C, TGFβ1, XRCC3, XRCC1, NAT2, and MMP-3 to increase susceptibility to oral submucous fibrosis, while SNPs related to p53, p73, MDM2, BCL2, CASP1, CASP10, MREIIA, PRKDC, POLRMT, and POLG2 increase susceptibility to oral leukoplakia in Indian patients [128–135]. However, reports of associations between SNPs of various genes and OPMD are not consistent in different studies and populations, and the evidence at present is limited [136].

Several lifestyles, environmental and psychological factors can also cause epigenetic changes due to dysregulation of DNA methylation, histone modifications, and miRNA expression which affect the pathogenesis and prognosis of oral cancer.

Since the epigenetic mechanisms are reversible, the aberrant signatures can be modified by exogenous epigenetic modifiers and can serve as therapeutic adjuncts along with traditional therapies in the future [137].

2.7.5 Immunosuppression

The concept of immunosurveillance and immunoediting act as important mechanisms in the development of cancer. There is evidence that the immune system which is supposed to suppress and eliminate malignant cells may themselves be suppressed by the function of tumor cells in oral cancers. The functional deficiencies of the tumor-associated immune cells in oral cancer are mediated by both soluble and contact-mediated immunosuppression by NFkB and STAT3 [138]. Human conditions with primary immune deficiencies have a 1.6- to 2.3-fold global increase in cancers as reported by large epidemiological studies [139]. Patients with acquired immune deficiencies like HIV infection are also at an increased risk for renal, oral, breast, skin, and colorectal cancer [140]. There is also an increased risk of HPV-associated anal cancer in HIV-positive men and cervical cancer in HIV-positive women [141]. An increased risk of SCC of the buccal cavity and skin after solid organ transplant and hematopoietic stem cell transplantation have been reported. They have been attributed to the chemotherapy, total body irradiation, and immunosuppressive therapy of cGVHD that occurs in these patients. Drugs like cyclophosphamide, cyclosporine, methotrexate, azathioprine, and corticosteroids have been implicated [142]. In hematopoietic stem cell transplant patients, the risk of oral cancer is 10 times more with a latency period of 5–9 years with tongue being the most common site [143]. The malignant potential of oral lichenoid reactions occurring in cGVHD may also be a cause for increased risk of oral cancer in such patients [143]. Patients with inflammatory bowel disease are also believed to be at high risk for developing oral cancer and precancerous lesions. The risk of oral cancer in IBD is estimated to be 1.08–1.78. Lip, buccal mucosa, gingiva, and tongue are the common sites, and the risk factors like immunosuppressive therapy with azathiopurines, presence of precancerous lesions, and HPV infection have been implicated [144, 145]. A case of oropharyngeal carcinoma arising after treatment with methotrexate and etanercept for treatment in rheumatoid arthritis has also been reported [146]. Regular screening of oral cavity in patients before, during, and after immunosuppressive therapy especially in high-risk groups is recommended.

2.8 Protective Factors

2.8.1 Diet

Nutritional deficiency has been considered responsible in about 20% of cancers in low-income countries [147]. Consuming more fruits, green and yellow vegetables, and micronutrients like vitamin C, folate, and selenium [148] is associated with a lower risk of cancer; however, this might be confounded by smoking and alcohol consumption, making the evidence suggestive of only a protective effect [149, 150]. The oral cancer risk appears to be reduced by about 50% by the addition of one serving of fruit and vegetables per day [151–153]. Two Indian case–control studies have reported a protective effect of reduction in the oral cancer risk on increasing the intake of fruits and vegetables [107, 154]; Gupta et al. observed such a protective effect on consuming yellow, cruciferous vegetables, and citrus fruits more than once a week [107]. A population-based case–control study in Kerala showed an inverse relationship of the intake of fruits and, to some extent of vegetables, and their correlate, beta-carotene, with OPMDs, in a dose–response fashion [155].

Vegetarians tend to consume more fruits and vegetables than comparable non-vegetarians, owing to which the overall cancer risk of the former (and vegans) might be slightly lower; however, there is no conclusive information available regarding individual cancers [156]. Four Indian case–control studies have reported an association between a nonvegetarian diet and increased oral cancer risk [107, 154, 157, 158]; Gupta et al. observed an increased oral cancer risk on consuming red meat more than once a week [107].

2.8.2 Vitamins and Minerals

It has been difficult to establish the link between increased susceptibility to certain types of cancer due to deficiencies of vitamins and minerals [159]. A case–control study in Andhra Pradesh has reported an increased risk of oral cancer due to low intake of thiamine, riboflavin, folic acid, vitamin C, magnesium, iron, and copper [154]. Two other case–control studies have shown protective linear effects or risk reduction in OPMDs, on the adequate intake of zinc, calcium, riboflavin, iron, and vitamin [155, 160]. Consuming high doses of supplements of vitamins or minerals has not shown reduction in the risk of cancer in well-nourished populations, but some have shown an increased risk (e.g., increased risk of lung cancer due to a high-dose β-carotene) [161]. Hence, it has been suggested to use vitamin and mineral supplements in moderation for cancer prevention [159].

2.8.3 Tea and Coffee

Tea and coffee [162] are among the most consumed beverages worldwide, with the former being the second most popular after water [163] (>2 billion cups consumed daily) [164]. Both beverages have numerous health benefits associated with their consumption as described below.

Tea is usually a concoction of water and cured leaves or leaf buds of the plant *Camellia sinensis* [163]. A recent umbrella review of 96 meta-analyses reported reduced risks of type 2 diabetes mellitus (DM), coronary artery disease, stroke, cardiac death, [162] total mortality, in addition to beneficial associations for several cancers, cognitive, skeletal, and maternal outcomes [165]. A phase II randomized controlled trial also reported a possible role of green tea extracts in suppressing OPMDs by reducing angiogenic stimulus, with higher doses showing improvement in short-term (12 week) outcome [166]. Coffee, an infusion of ground, roasted coffee beans [167], has been associated with a probable decreased risk of multiple cancers such as breast, colorectal, colon, endometrial, and prostate, type 2 DM, cardiovascular disease and mortality, and Parkinson's disease [162]. A systematic review of 17 studies found a protective effect of coffee on oral cavity cancers (pooled OR [95% CI] = 0.82 [0.58–1.16], $p = 0.257$) and for pharyngeal cancers (pooled OR [95% CI] = 0.72 [0.54–0.95]; $p = 0.01$) [168].

2.9 Conclusion

Oral cancer is a highly complex multifactorial disease where traditional major risk factors tobacco, alcohol, areca nut consumption play a major role in its causation. Several other factors like socioeconomic status, microbial infections, nutritional deficiencies, immunosuppression, and genetic factors could contribute in a multistep process to carcinogenesis. While some of these factors are inherent and nonmodifiable, majority of them are modifiable, and efforts should be made toward early identification and elimination of these factors for risk-reduction and primary prevention of oral cancer.

References

1. Sung H, Ferlay J, Siegel RL, Laversanne M, Soerjomataram I, Jemal A et al (2021) Global cancer statistics 2020: GLOBOCAN estimates of incidence and mortality worldwide for 36 cancers in 185 countries. CA Cancer J Clin 68:394. https://acsjournals.onlinelibrary.wiley.com/doi/abs/10.3322/caac.21660. Accessed 16 Feb 2021
2. Warnakulasuriya S, Greenspan JS (2020) Epidemiology of oral and oropharyngeal cancers. Springer, Cham. https://link.springer.com/chapter/10.1007/978-3-030-32316-5_2. Accessed 16 Feb 2021
3. WHO. IARC (n.d.) Cancer Today. http://gco.iarc.fr/today/home. Accessed 17 Feb 2021
4. Johnson DE, Burtness B, Leemans CR, Lui VWY, Bauman JE, Grandis JR (2020) Head, and neck squamous cell carcinoma. Nat Rev Dis Primer 6(1):1–22
5. Laronde DM (2008) Oral cancer: just the facts. J Can Dent Assoc 74(3):4
6. Dovjak M, Kukec A (2019) Identification of health risk factors and their parameters. In: Dovjak M, Kukec A (eds) Creating healthy and sustainable buildings: an assessment of health risk factors. Springer International Publishing, Cham, pp 83–120. https://doi.org/10.1007/978-3-030-19412-3_3. Accessed 16 Feb 2021
7. McDowell JD (2006) An overview of epidemiology and common risk factors for oral squamous cell carcinoma. Otolaryngol Clin N Am 39(2):277–294
8. IARC Working Group on the Evaluation of Carcinogenic Risks to Humans (2012) Personal habits and indoor combustions. A review of human carcinogens. IARC Monogr Eval Carcinog Risks Hum 100(Pt E):1–538
9. WHO (n.d.) Tobacco. https://www.who.int/news-room/fact-sheets/detail/tobacco. Accessed 16 Feb 2021
10. TISS (n.d.) Outcomes/publications. Global adult tobacco survey 2, India 2016-17 - research project. https://www.tiss.edu/view/6/mumbai-campus/school-of-health-systems-studies/

global-adult-tobacco-survey-2-india-2016-17/outcomespublications-3/. Accessed 17 Feb 2021
11. NHM (n.d.) Index of /NTCP/surveys-reports-publications. https://nhm.gov.in/NTCP/Surveys-Reports-Publications/. Accessed 17 Feb 2021
12. Campaign for Tobacco Free Kids. The toll of tobacco in India. Available from: https://www.tobaccofreekids.org/problem/toll-global/asia/india. Accessed 14 Mar 2022
13. World Health Organisation India. Tobacco. Available from: https://www.who.int/india/health-topics/tobacco#:~:text=Tobacco%20use%20is%20a%20major,1.35%20million%20deaths%20every%20year. Accessed 14 Mar 2022
14. Gupta P, Asma S (2008) Bidi smoking and public health. Ministry of Health and Family Welfare, New Delhi
15. IARC (n.d.) Report of the Advisory Group to recommend priorities for the IARC monographs during 2020–2024 – IARC. https://www.iarc.who.int/news-events/report-of-the-advisory-group-to-recommend-priorities-for-the-iarc-monographs-during-2020-2024/. Accessed 16 Feb 2021
16. Bitzer ZT, Goel R, Reilly SM, Elias RJ, Silakov A, Foulds J et al (2018) Effect of flavoring chemicals on free radical formation in electronic cigarette aerosols. Free Radic Biol Med 120: 72–79
17. Chakma JK, Dhaliwal RS, Mehrotra R (2019) White Paper on Electronic Nicotine Delivery System. Indian J Med Res 149(5):574–583
18. Bharath TS, Kumar NGR, Nagaraja A, Saraswathi T, Babu GS, Raju PR (2015) Palatal changes of reverse smokers in a rural coastal Andhra population with review of literature. J Oral Maxillofac Pathol 19(2):182–187
19. Gavarasana S, Susarla MD (1989) Palatal mucosal changes among reverse smokers in an Indian village. Jpn J Cancer Res GANN 80(3):209–211
20. Mehta FS, Jalnawalla PN, Daftary DK, Gupta PC, Pindborg JJ (1977) Reverse smoking in Andhra Pradesh, India: variability of clinical and histologic appearances of palatal changes. Int J Oral Surg 6(2):75–83
21. IARC Working Group on the Evaluation of Carcinogenic Risks to Humans (2004) Tobacco smoke and involuntary smoking. IARC Monogr Eval Carcinog Risks Hum 83:1–1438
22. Hecht SS (2008) Progress and challenges in selected areas of tobacco carcinogenesis. Chem Res Toxicol 21(1):160–171
23. Hecht SS, Carmella SG, Le K-A, Murphy SE, Boettcher AJ, Le C et al (2006) 4-(methylnitrosamino)-1-(3-pyridyl)-1-butanol and its glucuronides in the urine of infants exposed to environmental tobacco smoke. Cancer Epidemiol Biomark Prev 15(5):988–992
24. Suwan-ampai P, Navas-Acien A, Strickland PT, Agnew J (2009) Involuntary tobacco smoke exposure and urinary levels of polycyclic aromatic hydrocarbons in the United States, 1999 to 2002. Cancer Epidemiol Biomark Prev 18(3):884–893
25. Division of Cancer Control and Population Sciences (DCCPS) (n.d.) Smokeless tobacco and public health: a global perspective. https://cancercontrol.cancer.gov/brp/tcrb/smokeless-tobacco. Accessed 17 Feb 2021
26. Wyss A, Hashibe M, Chuang S-C, Lee Y-CA, Zhang Z-F, Yu G-P et al (2013) Cigarette, cigar, and pipe smoking and the risk of head and neck cancers: pooled analysis in the International Head and Neck Cancer Epidemiology Consortium. Am J Epidemiol 178(5):679–690
27. Znaor A, Brennan P, Gajalakshmi V, Mathew A, Shanta V, Varghese C et al (2003) Independent and combined effects of tobacco smoking, chewing and alcohol drinking on the risk of oral, pharyngeal, and esophageal cancers in Indian men. Int J Cancer 105(5):681–686
28. Rahman M, Sakamoto J, Fukui T (2003) Bidi smoking and oral cancer: a meta-analysis. Int J Cancer 106(4):600–604
29. Muwonge R, Ramadas K, Sankila R, Thara S, Thomas G, Vinoda J et al (2008) Role of tobacco smoking, chewing and alcohol drinking in the risk of oral cancer in Trivandrum, India: a nested case-control design using incident cancer cases. Oral Oncol 44(5):446–454

30. Hashibe M, Li Q, Chen C-J, Hsu W-L, Lou P-J, Zhu C et al (2019) Involuntary smoking and the risk of head and neck cancer in an East Asian population. Cancer Epidemiol 59:173–177
31. Lee Y-CA, Marron M, Benhamou S, Bouchardy C, Ahrens W, Pohlabeln H et al (2009) Active and involuntary tobacco smoking and upper aerodigestive tract cancer risks in a multicenter case-control study. Cancer Epidemiol Biomark Prev 18(12):3353–3361
32. Asthana S, Labani S, Kailash U, Sinha DN, Mehrotra R (2019) Association of Smokeless Tobacco Use and Oral Cancer: a systematic global review and meta-analysis. Nicotine Tob Res 21(9):1162–1171
33. Guha N, Warnakulasuriya S, Vlaanderen J, Straif K (2014) Betel quid chewing and the risk of oral and oropharyngeal cancers: a meta-analysis with implications for cancer control. Int J Cancer 135(6):1433–1443
34. Goldstein BY, Chang S-C, Hashibe M, La Vecchia C, Zhang Z-F (2010) Alcohol consumption and cancers of the oral cavity and pharynx from 1988 to 2009: an update. Eur J Cancer Prev 19(6):431–465
35. Tramacere I, Negri E, Bagnardi V, Garavello W, Rota M, Scotti L et al (2010) A meta-analysis of alcohol drinking and oral and pharyngeal cancers. Part 1: overall results and dose-risk relation. Oral Oncol 46(7):497–503
36. Singhavi HR, Singh A, Bhattacharjee A, Talole S, Dikshit R, Chaturvedi P (2020) Alcohol, and cancer risk: a systematic review and meta-analysis of prospective Indian studies. Indian J Public Health 64(2):186–190
37. Warnakulasuriya S, Trivedy C, Peters TJ (2002) Areca nut use: an independent risk factor for oral cancer. BMJ 324(7341):799–800
38. Warnakulasuriya S, Peters TJ (2002) Introduction: biology, medical and socio-economic aspects of areca nut use. Addict Biol 7(1):75–76
39. IARC (n.d.) Betel-quid and areca-nut chewing and some areca-nut-derived nitrosamines. https://publications.iarc.fr/Book-And-Report-Series/Iarc-Monographs-On-The-Identification-Of-Carcinogenic-Hazards-To-Humans/Betel-quid-And-Areca-nut-Chewing-And-Some-Areca-nut-derived-Nitrosamines-2004. Accessed 16 Feb 2021
40. Patidar KA, Parwani R, Wanjari SP, Patidar AP (2015) Various terminologies associated with areca nut and tobacco chewing: a review. J Oral Maxillofac Pathol 19(1):69–76
41. Chandra PS, Mulla U (2007) Areca nut: the hidden Indian "gateway" to future tobacco use and oral cancers among youth. Indian J Med Sci 61(6):319–321
42. Gupta PC, Ray CS, Papke RL, Stepanov I, Khariwala SS, Chaturvedi P et al (2018) Perspectives on areca nut with some global implications: symposium report. Transl Res Oral Oncol 3:2057178X18814068
43. Marques MM, Beland FA, Lachenmeier DW, Phillips DH, Chung F-L, Dorman DC et al (2021) Carcinogenicity of acrolein, crotonaldehyde, and arecoline. Lancet Oncol 22(1):19–20
44. WHO (n.d.) Alcohol. https://www.who.int/westernpacific/health-topics/alcohol. Accessed 18 Feb 2021
45. GHO (n.d.) World Health Statistics data visualizations dashboard. Harmful use of alcohol. World Health Organization; https://apps.who.int/gho/data/node.sdg.3-5-viz?lang=en. Accessed 16 Feb 2021
46. Organisation mondiale de la santé, Organisation mondiale de la santé, Management of Substance Abuse Team (2018) Global status report on alcohol and health 2018
47. Gururaj G, Gautham MS, Arvind BA (2021) Alcohol consumption in India: a rising burden and a fractured response. Drug Alcohol Rev 40:368. https://onlinelibrary.wiley.com/doi/abs/10.1111/dar.13179. Accessed 16 Feb 2021
48. Google My Maps (n.d.) National Drug Dependence Treatment Centre. https://www.google.com/maps/d/viewer?mid=1VvJZVcy9uc2XyidRQrGh6V80u8A. Accessed 18 Feb 2021
49. RCHIIPS (n.d.) National Family Health Survey. http://rchiips.org/nfhs/. Accessed 16 Feb 2021

50. Stewart BW, Wild CP (2014) World Cancer Report 2014. https://publications.iarc.fr/Non-Series-Publications/World-Cancer-Reports/World-Cancer-Report-2014. Accessed 18 Feb 2021
51. Pelucchi C, Tramacere I, Boffetta P, Negri E, La Vecchia C (2011) Alcohol consumption, and cancer risk. Nutr Cancer 63(7):983–990
52. Reidy JT, McHugh EE, Stassen LFA (2011) A review of the role of alcohol in the pathogenesis of oral cancer and the link between alcohol-containing mouthrinses and oral cancer. J Irish Dent Assoc 57(4):200–202
53. Ogden GR (2005) Alcohol and oral cancer. Alcohol Fayettev N 35(3):169–173
54. Squier CA, Cox P, Hall BK (1986) Enhanced penetration of nitrosonornicotine across oral mucosa in the presence of ethanol. J Oral Pathol 15(5):276–279
55. Seitz HK, Stickel F (2007) Molecular mechanisms of alcohol-mediated carcinogenesis. Nat Rev Cancer 7(8):599–612
56. Warnakulasuriya S, Parkkila S, Nagao T, Preedy VR, Pasanen M, Koivisto H et al (2008) Demonstration of ethanol-induced protein adducts in oral leukoplakia (pre-cancer) and cancer. J Oral Pathol Med 37(3):157–165
57. Mello FW, Melo G, Pasetto JJ, Silva CAB, Warnakulasuriya S, Rivero ERC (2019) The synergistic effect of tobacco and alcohol consumption on oral squamous cell carcinoma: a systematic review and meta-analysis. Clin Oral Investig 23(7):2849–2859
58. Conway DI, Petticrew M, Marlborough H, Berthiller J, Hashibe M, Macpherson LMD (2008) Socioeconomic inequalities and oral cancer risk: a systematic review and meta-analysis of case-control studies. Int J Cancer 122(12):2811–2819
59. Warnakulasuriya S (2009) Significant oral cancer risk associated with low socioeconomic status. Evid Based Dent 10(1):4–5
60. Johnson S, McDonald JT, Corsten M (2012) Oral cancer screening and socioeconomic status. J Otolaryngol Head Neck Surg 41(2):102–107
61. Auluck A, Walker BB, Hislop G, Lear SA, Schuurman N, Rosin M (2016) Socio-economic deprivation: a significant determinant affecting stage of oral cancer diagnosis and survival. BMC Cancer 16(1):569
62. Lohe VK (2017) Association of socioeconomic risk factor with patients delay in presentation of oral squamous cell carcinoma. http://www.journaldmims.com/article.asp?issn=0974-3901;year=2017;volume=12;issue=1;spage=75;epage=78;aulast=Lohe. Accessed 16 Jan 2021
63. Lee C-C, Chien S-H, Hung S-K, Yang W-Z, Su Y-C (2012) Effect of individual and neighborhood socioeconomic status on oral cancer survival. Oral Oncol 48(3):253–261
64. Beachler DC, D'Souza G (2013) Oral HPV infection and head and neck cancers in HIV-infected individuals. Curr Opin Oncol 25(5):503–510
65. Syrjänen S (2018) Oral manifestations of human papillomavirus infections. Eur J Oral Sci 126 (Suppl 1):49–66
66. International Agency for Research on Cancer (2012) Human papillomaviruses. Biological agents. IARC, Lyon. https://www.ncbi.nlm.nih.gov/books/NBK304347/. Accessed 1 Feb 2021
67. Yete S, D'Souza W, Saranath D (2018) High-risk human papillomavirus in oral cancer: clinical implications. Oncology 94(3):133–141
68. Irani S, Barati I, Badiei M (2020) Periodontitis, and oral cancer - current concepts of the etiopathogenesis. Oncol Rev 14:465. https://www.ncbi.nlm.nih.gov/pmc/articles/PMC7097927/. Accessed 7 Jan 2021
69. Nair D, Mair M, Singh A, D'Cruz A (2018) Prevalence and impact of human papillomavirus on head and neck cancers: review of Indian studies. Indian J Surg Oncol 9(4):568–575
70. de la Cour CD, Sperling CD, Belmonte F, Syrjänen S, Kjaer SK (2021) Human papillomavirus prevalence in oral potentially malignant disorders: systematic review and meta-analysis. Oral Dis 27:431
71. Young LS, Yap LF, Murray PG (2016) Epstein–Barr virus: more than 50 years old and still providing surprises. Nat Rev Cancer 16(12):789–802

72. de Lima MAP, Teodoro IPP, de Galiza LE, Filho PHBM, Marques FM, Pinheiro Junior RFF et al (2019) Association between Epstein-Barr virus and oral carcinoma: a systematic review with meta-analysis. Crit Rev Oncog 24(4):349–368
73. Guidry JT, Birdwell CE, Scott RS (2018) Epstein–Barr virus in the pathogenesis of oral cancers. Oral Dis 24(4):497–508
74. She Y, Nong X, Zhang M, Wang M (2017) Epstein-Barr virus infection and oral squamous cell carcinoma risk: a meta-analysis. PLoS One 12:e0186860. https://www.ncbi.nlm.nih.gov/pmc/articles/PMC5655447/. Accessed 23 Jan 2021
75. Sivakumar S, Gupta AA, Rosdy NMMNM, Venkiteswaran A, Raj AT, Awan KH (2020) Assessing the potential association between Epstein-Barr virus and oral squamous cell carcinoma: a systematic review and meta- analysis. Transl. Cancer Res 9(4):3092. http://tcr.amegroups.com/article/view/37352. Accessed 2 Jan 2021
76. Sankari SL, Gayathri K, Balachander N, Malathi L (2015) Candida in potentially malignant oral disorders. J Pharm Bioallied Sci 7(Suppl 1):S162–S164
77. Shukla K, Vun I, Lov I, Laparidis G, McCamley C, Ariyawardana A (2019) Role of Candida infection in the malignant transformation of oral leukoplakia: a systematic review of observational studies. Transl Res Oral Oncol 4:2057178X19828229
78. Ramirez-Garcia A, Rementeria A, Aguirre-Urizar JM, Moragues MD, Antoran A, Pellon A et al (2016) Candida albicans and cancer: can this yeast induce cancer development or progression? Crit Rev Microbiol 42(2):181–193
79. Arzmi MH, Dashper S, McCullough M (2019) Polymicrobial interactions of Candida albicans and its role in oral carcinogenesis. J Oral Pathol Med 48(7):546–551
80. Satiman EAFEN, Ahmad H, Ramzi AB, Wahab RA, Kaderi MA, Harun WHAW et al (2020) The role of Candida albicans candidalysin ECE1 gene in oral carcinogenesis. J Oral Pathol Med 49(9):835–841
81. Gupta K, Metgud R (2013) Evidence suggesting involvement of viruses in oral squamous cell carcinoma. Pathol Res Int 2013:e642496. https://www.hindawi.com/journals/pri/2013/642496/. Accessed 25 Jan 2021
82. Foy J-P, Bertolus C, Boutolleau D, Agut H, Gessain A, Herceg Z et al (2020) Arguments to support a viral origin of oral squamous cell carcinoma in non-smoker and non-drinker patients. Front Oncol 10:822
83. Jain M (2016) Assessment of correlation of herpes simplex virus-1 with oral cancer and precancer - a comparative study. J Clin Diagn Res 10(8):ZC14–ZC17
84. Cox M, Maitland N, Scully C (1993) Human herpes simplex-1 and papillomavirus type 16 homologous DNA sequences in normal, potentially malignant, and malignant oral mucosa. Eur J Cancer B Oral Oncol 29B(3):215–219
85. Yildirim B, Sengüven B, Demir C (2011) Prevalence of herpes simplex, Epstein Barr, and human papilloma viruses in oral lichen planus. Med Oral Patol Oral Cir Bucal 16(2):e170–e174
86. Shillitoe EJ, Gilchrist E, Pellenz C, Murrah V (1999) Effects of herpes simplex virus on human oral cancer cells, and potential use of mutant viruses in therapy of oral cancer. Oral Oncol 35(3):326–332
87. Yura Y (2017) Presage of oncolytic virotherapy for oral cancer with herpes simplex virus. Jpn Dent Sci Rev 53(2):53–60
88. Kulkarni V, Uttamani JR, Bhatavadekar NB (2016) Comparison of clinical periodontal status among habitual smokeless-tobacco users and cigarette smokers. Int Dent J 66(1):29–35
89. Bhandarkar GP, Shetty KV, Ashaya, Jha K, Arati K, Thomas T (2020) Correlation of periodontal parameters to various types of smokeless tobacco in tobacco pouch keratosis patients: a cross-sectional study. J Cancer Res Ther 16(3):463–469
90. Giovannoni ML, Valdivia-Gandur I, Lozano de Luaces V, Varela Véliz H, Balasubbaiah Y, Chimenos-Küstner E (2018) Betel and tobacco chewing habit and its relation to risk factors for periodontal disease. Oral Dis 24(5):829–839

91. Wang J, Lv J, Wang W, Jiang X (2016) Alcohol consumption and risk of periodontitis: a meta-analysis. J Clin Periodontol 43(7):572–583
92. Holliday RS, Campbell J, Preshaw PM (2019) Effect of nicotine on human gingival, periodontal ligament and oral epithelial cells. A systematic review of the literature. J Dent 86:81–88
93. Zhang Y, He J, He B, Huang R, Li M (2019) Effect of tobacco on periodontal disease and oral cancer. Tob Induc Dis 17:40
94. Sadighi Shamami M, Sadighi Shamami M, Amini S (2011) Periodontal disease and tooth loss as risks for cancer: a systematic review of the literature. Iran J Cancer Prev 4(4):189–198
95. Dayama A, Srivastava V, Shukla M, Singh R, Pandey M (2011) Helicobacter pylori and oral cancer: possible association in a preliminary case control study. Asian Pac J Cancer Prev 12(5):1333–1336
96. Ibáñez L, de Mendoza I, Maritxalar Mendia X, García de la Fuente AM, Quindós Andrés G, Aguirre Urizar JM (2020) Role of porphyromonas gingivalis in oral squamous cell carcinoma development: a systematic review. J Periodontal Res 55(1):13–22
97. Gopinath D, Menon RK, Veettil SK, George Botelho M, Johnson NW (2020) Periodontal diseases as putative risk factors for head and neck cancer: systematic review and meta-analysis. Cancers 12:1893. https://www.ncbi.nlm.nih.gov/pmc/articles/PMC7409086/. Accessed 7 Jan 2021
98. La Rosa GRM, Gattuso G, Pedullà E, Rapisarda E, Nicolosi D, Salmeri M (2020) Association of oral dysbiosis with oral cancer development. Oncol Lett 19(4):3045–3058
99. Warnakulasuriya S (2009) Causes of oral cancer--an appraisal of controversies. Br Dent J 207(10):471–475
100. Piemonte ED, Lazos J (2018) Evidence for chronic mechanical irritation as a causative factor in oral cancer: application of Bradford Hill criteria. Transl Res Oral Oncol 3: 2057178X18791034
101. Singhvi HR, Malik A, Chaturvedi P (2017) The role of chronic mucosal trauma in oral cancer: a review of literature. Indian J Med Paediatr Oncol 38(1):44–50
102. Rangaswamy S, Chikkalingaiah RG, Sanjeevarayappa PN, Govindraju P (2019) Carcinoma arising in the background of oral submucous fibrosis. Ann Maxillofac Surg 9(2):247–252
103. Piemonte E, Lazos J, Belardinelli P, Secchi D, Brunotto M, Lanfranchi-Tizeira H (2018) Oral cancer associated with chronic mechanical irritation of the oral mucosa. Med Oral Patol Oral Cir Bucal 23(2):e151–e160
104. Gandini S, Negri E, Boffetta P, La Vecchia C, Boyle P (2012) Mouthwash, and oral cancer risk quantitative meta-analysis of epidemiologic studies. Ann Agric Environ Med 19(2):173–180
105. Ahrens W, Pohlabeln H, Foraita R, Nelis M, Lagiou P, Lagiou A et al (2014) Oral health, dental care, and mouthwash associated with upper aerodigestive tract cancer risk in Europe: the ARCAGE study. Oral Oncol 50(6):616–625
106. Patel A, Pathak Y, Patel J, Sutariya V (2018) Role of nutritional factors in pathogenesis of cancer. Food Qual Saf 2(1):27–36
107. Gupta B, Bray F, Kumar N, Johnson NW (2017) Associations between oral hygiene habits, diet, tobacco and alcohol and risk of oral cancer: a case-control study from India. Cancer Epidemiol 51:7–14
108. Loomis D, Guyton KZ, Grosse Y, Lauby-Secretan B, El Ghissassi F, Bouvard V et al (2016) Carcinogenicity of drinking coffee, mate, and very hot beverages. Lancet Oncol 17(7): 877–878
109. IARC (2018) Drinking coffee, mate, and very hot beverages. https://publications.iarc.fr/Book-And-Report-Series/Iarc-Monographs-On-The-Identification-Of-Carcinogenic-Hazards-To-Humans/Drinking-Coffee-Mate-And-Very-Hot-Beverages-2018. Accessed 16 Feb 2021
110. Dasanayake AP, Silverman AJ, Warnakulasuriya S (2010) Maté drinking and oral and oro-pharyngeal cancer: a systematic review and meta-analysis. Oral Oncol 46(2):82–86

111. Mello FW, Scotti FM, Melo G, Warnakulasuriya S, Guerra ENS, Rivero ERC (2018) Maté consumption association with upper aerodigestive tract cancers: a systematic review and meta-analysis. Oral Oncol 82:37–47
112. IARC (1991) Occupational exposures in insecticide application, and some pesticides. https://publications.iarc.fr/Book-And-Report-Series/Iarc-Monographs-On-The-Identification-Of-Carcinogenic-Hazards-To-Humans/Occupational-Exposures-In-Insecticide-Application-And-Some-Pesticides-1991. Accessed 16 Feb 2021
113. IARC (2012) Chemical agents and related occupations. https://publications.iarc.fr/Book-And-Report-Series/Iarc-Monographs-On-The-Identification-Of-Carcinogenic-Hazards-To-Humans/Chemical-Agents-And-Related-Occupations-2012. Accessed 16 Feb 2021
114. CDC (2018) Bloodborne infectious diseases - occupations affected by bloodborne infectious diseases - NIOSH Workplace Safety and Health Topic. https://www.cdc.gov/niosh/topics/bbp/occupations.html. Accessed 16 Feb 2021
115. Awan KH, Hegde R, Cheever VJ, Carroll W, Khan S, Patil S et al (2018) Oral and pharyngeal cancer risk associated with occupational carcinogenic substances: systematic review. Head Neck 40(12):2724–2732
116. Wu W-T, Lin Y-J, Li C-Y, Tsai P-J, Yang C-Y, Liou S-H et al (2015) Cancer attributable to asbestos exposure in shipbreaking workers: a matched-cohort study. PLoS One 10(7):e0133128
117. Schildt EB, Eriksson M, Hardell L, Magnuson A (1999) Occupational exposures as risk factors for oral cancer evaluated in a Swedish case-control study. Oncol Rep 6(2):317–320
118. Laforest L, Luce D, Goldberg P, Bégin D, Gérin M, Demers PA et al (2000) Laryngeal and hypopharyngeal cancers and occupational exposure to formaldehyde and various dusts: a case-control study in France. Occup Environ Med 57(11):767–773
119. INCHEM (1989) Occupational exposures in paint manufacture and painting (IARC summary & evaluation), vol 47. http://www.inchem.org/documents/iarc/vol47/47-13.html. Accessed 16 Feb 2021
120. Merletti F, Boffetta P, Ferro G, Pisani P, Terracini B (1991) Occupation and cancer of the oral cavity or oropharynx in Turin, Italy. Scand J Work Environ Health 17(4):248–254
121. Sarode GS, Batra A, Sarode SC, Yerawadekar S, Patil S (2016) Oral cancer-related inherited cancer syndromes: a comprehensive review. J Contemp Dent Pract 17(6):504–510
122. Wright JT (2010) Oral manifestations in the epidermolysis bullosa spectrum. Dermatol Clin 28(1):159–164
123. Trott KE, Briddell JW, Corao-Uribe D, Powell J, Seecof OM, Levy C et al (2019) Dyskeratosis congenita and oral cavity squamous cell carcinoma: report of a case and literature review. J Pediatr Hematol Oncol 41(6):501–503
124. Ankathil R, Mathew A, Joseph F, Nair MK (1996) Is oral cancer susceptibility inherited? Report of five oral cancer families. Eur J Cancer B Oral Oncol 32(1):63–67
125. Sharma V, Nandan A, Sharma AK, Singh H, Bharadwaj M, Sinha DN et al (2017) Signature of genetic associations in oral cancer. Tumor Biol 39(10):1010428317725923
126. Tsantoulis PK, Kastrinakis NG, Tourvas AD, Laskaris G, Gorgoulis VG (2007) Advances in the biology of oral cancer. Oral Oncol 43(6):523–534
127. Multani S, Saranath D (2016) Genotypic distribution of single nucleotide polymorphisms in oral cancer: global scene. Tumour Biol 37(11):14501–14512
128. Gupta SR, Sharma A, Gupta N, Mani K (2020) Single nucleotide polymorphisms and serologic levels of hypoxia-inducible factor1 α and vascular endothelial growth factor are associated with increased risk of oral submucous fibrosis in gutka users among a North Indian population. Oral Surg Oral Med Oral Pathol Oral Radiol 130(5):557–564
129. Rajendran R, Harish RK, Anil S, Vidyadharan R, Banerjee M (2010) Transforming growth factor-β-1 polymorphisms are infrequent but exist at selected loci in oral submucous fibrosis. Indian J Dent Res 21(3):413–419

130. Mukherjee S, Bhowmik AD, Roychoudhury P, Mukhopadhyay K, Ray JG, Chaudhuri K (2012) Association of XRCC1, XRCC3, and NAT2 polymorphisms with the risk of oral submucous fibrosis among eastern Indian population. J Oral Pathol Med 41(4):292–302
131. Chaudhary AK, Singh M, Bharti AC, Singh M, Shukla S, Singh AK et al (2010) Synergistic effect of stromelysin-1 (matrix metalloproteinase-3) promoter (-1171 5A->6A) polymorphism in oral submucous fibrosis and head and neck lesions. BMC Cancer 10:369
132. Misra C, Majumder M, Bajaj S, Ghosh S, Roy B, Roychoudhury S (2009) Polymorphisms at p53, p73, and MDM2 loci modulate the risk of tobacco associated leukoplakia and oral cancer. Mol Carcinog 48(9):790–800
133. Datta S, Ray A, Singh R, Mondal P, Basu A, De Sarkar N et al (2015) Sequence and expression variations in 23 genes involved in mitochondrial and non-mitochondrial apoptotic pathways and risk of oral leukoplakia and cancer. Mitochondrion 25:28–33
134. Mondal P, Datta S, Maiti GP, Baral A, Jha GN, Panda CK et al (2013) Comprehensive SNP scan of DNA repair and DNA damage response genes reveal multiple susceptibility loci conferring risk to tobacco associated leukoplakia and oral cancer. PLoS One 8(2):e56952
135. Datta S, Ray A, Roy R, Roy B (2016) Association of DNA sequence variation in mitochondrial DNA polymerase with mitochondrial DNA synthesis and risk of oral cancer. Gene 575(2 Pt 3):650–654
136. Shridhar K, Aggarwal A, Walia GK, Gulati S, Geetha AV, Prabhakaran D et al (2016) Single nucleotide polymorphisms as markers of genetic susceptibility for oral potentially malignant disorders risk: review of evidence to date. Oral Oncol 61:146–151
137. Irimie AI, Ciocan C, Gulei D, Mehterov N, Atanasov AG, Dudea D et al (2018) Current insights into oral cancer epigenetics. Int J Mol Sci 19(3):670
138. Jewett A, Head C, Cacalano NA (2006) Emerging mechanisms of immunosuppression in oral cancers. J Dent Res 85(12):1061–1073
139. Satgé D (2018) A tumor profile in primary immune deficiencies challenges the cancer immune surveillance concept. Front Immunol 9:1149. https://www.ncbi.nlm.nih.gov/pmc/articles/PMC5976747/. Accessed 15 Jan 2021
140. Chen C-H, Chung C-Y, Wang L-H, Lin C, Lin H-L, Lin H-C (2015) Risk of cancer among HIV-infected patients from a population-based nested case–control study: implications for cancer prevention. BMC Cancer 15(1):133
141. Palefsky J (2009) Human papillomavirus-related disease in people with HIV. Curr Opin HIV AIDS 4(1):52–56
142. Curtis RE, Metayer C, Rizzo JD, Socié G, Sobocinski KA, Flowers MED et al (2005) Impact of chronic GVHD therapy on the development of squamous-cell cancers after hematopoietic stem-cell transplantation: an international case-control study. Blood 105(10):3802–3811
143. Kruse AL, Grätz KW (2009) Oral carcinoma after hematopoietic stem cell transplantation – a new classification based on a literature review over 30 years. Head Neck Oncol 1(1):29
144. Katsanos KH, Roda G, Brygo A, Delaporte E, Colombel J-F (2015) Oral cancer and oral precancerous lesions in inflammatory bowel diseases: a systematic review. J Crohns Colitis 9(11):1043–1052
145. Li ACY, Warnakulasuriya S, Thompson RPH (2003) Neoplasia of the tongue in a patient with Crohn's disease treated with azathioprine: case report. Eur J Gastroenterol Hepatol 15(2): 185–187
146. Chainani-Wu N, Chang C, Gross AJ, Yom SS, Silverman S (2014) Oropharyngeal carcinoma arising after methotrexate and etanercept therapy for rheumatoid arthritis. Oral Surg Oral Med Oral Pathol Oral Radiol 117(3):e261–e263
147. Key TJ, Schatzkin A, Willett WC, Allen NE, Spencer EA, Travis RC (2004) Diet, nutrition, and the prevention of cancer. Public Health Nutr 7(1a):187–200
148. Chen Q, Lin LS, Chen L, Lin J, Ding Y, Bao XD, et al (2019) [Relationship between selenium and the risk for oral cancer: a case-control study]. Zhonghua Liu Xing Bing Xue Za Zhi Zhonghua Liuxingbingxue Zazhi 40(7):810–814

149. World Cancer Research Fund (2018) Diet and cancer. https://www.wcrf.org/dietandcancer. Accessed 16 Feb 2021
150. Adami H-O, Hunter DJ, Lagiou P, Mucci L (eds) (2018) Textbook of cancer epidemiology, 3rd edn. Oxford University Press, Oxford; New York, NY. 752 p
151. la Vecchia C, Negri E, d'Avanzo B, Boyle P, Franceschi S (1991) Dietary indicators of oral and pharyngeal cancer. Int J Epidemiol 20(1):39–44
152. Zheng W, Blot WJ, Diamond EL, Norkus EP, Spate V, Morris JS et al (1993) Serum micronutrients and the subsequent risk of oral and pharyngeal cancer. Cancer Res 53(4): 795–798
153. Wilson M (ed) (2009) Food constituents and oral health: current status and future prospects, 1st edn. Woodhead Publishing, New Delhi. 568 p
154. Prasad MP, Krishna TP, Pasricha S, Quereshi MA, Krishnaswamy K (1995) Diet and oral cancer - a case control study. Asia Pac J Clin Nutr 4(2):259–264
155. Gupta PC, Hebert JR, Bhonsle RB, Murti PR, Mehta H, Mehta FS (1999) Influence of dietary factors on oral precancerous lesions in a population-based case-control study in Kerala, India. Cancer 85(9):1885–1893
156. Appleby PN, Key TJ (2016) The long-term health of vegetarians and vegans. Proc Nutr Soc 75(3):287–293
157. Subapriya R, Thangavelu A, Mathavan B, Ramachandran CR, Nagini S (2007) Assessment of risk factors for oral squamous cell carcinoma in Chidambaram, Southern India: a case-control study. Eur J Cancer Prev 16(3):251–256
158. Gangane N, Chawla S, Anshu N, Subodh A, Gupta SS, Sharma SM (2007) Reassessment of risk factors for oral cancer. Asian Pac J Cancer Prev 8(2):243–248
159. Key TJ, Bradbury KE, Perez-Cornago A, Sinha R, Tsilidis KK, Tsugane S (2020) Diet, nutrition, and cancer risk: what do we know and what is the way forward? BMJ 368:m511
160. Hebert JR, Gupta PC, Bhonsle RB, Mehta H, Zheng W, Sanderson M et al (2002) Dietary exposures and oral precancerous lesions in Srikakulam District, Andhra Pradesh, India. Public Health Nutr 5(2):303–312
161. Alpha-Tocopherol, Beta Carotene Cancer Prevention Study Group (1994) The effect of vitamin E and beta carotene on the incidence of lung cancer and other cancers in male smokers. N Engl J Med 330(15):1029–1035
162. Grosso G, Godos J, Galvano F, Giovannucci EL (2017) Coffee, caffeine, and health outcomes: an umbrella review. Annu Rev Nutr 37:131–156
163. Kuriyama S, Shimazu T, Ohmori K, Kikuchi N, Nakaya N, Nishino Y et al (2006) green tea consumption and mortality due to cardiovascular disease, cancer, and all causes in Japan: the Ohsaki study. JAMA 296(10):1255–1265
164. Brody H (2019) Tea. Nature 566(7742):S1
165. Yi M, Wu X, Zhuang W, Xia L, Chen Y, Zhao R et al (2019) Tea consumption and health outcomes: umbrella review of meta-analyses of observational studies in humans. Mol Nutr Food Res 63(16):e1900389
166. Tsao AS, Liu D, Martin J, Tang X, Lee JJ, El-Naggar AK et al (2009) Phase II randomized, placebo-controlled trial of green tea extract in patients with high-risk oral premalignant lesions. Cancer Prev Res (Phila) 2(11):931–941
167. Higdon JV, Frei B (2006) Coffee, and health: a review of recent human research. Crit Rev Food Sci Nutr 46(2):101–123
168. Miranda J, Monteiro L, Albuquerque R, Pacheco J-J, Khan Z, Lopez-Lopez J et al (2017) Coffee is protective against oral and pharyngeal cancer: a systematic review and meta-analysis. Med Oral Patol Oral Cir Bucal 22(5):e554–e561

Molecular Carcinopathogenesis in Oral Squamous Cell Carcinoma

3

Mukund Namdev Sable and Shubhada V. Kane

Abstract

Oral squamous cell carcinoma (OSCC) is a heterogeneous and multifactorial disease with generally an aggressive behavior. The molecular carcinogenesis of OSCC is a multistep process in which multiple genetic alterations bring functional changes to the oncogenes and tumor suppressor genes. Various molecular events can occur singly or in combinations, leading to changes in phenotypes of epithelial cells resulting in a series of mechanisms, viz. increased cell proliferation, epithelial-mesenchymal transition, increased loss of cell cohesion, and infiltrative nature with or without metastatic potential.

Recent advances in molecular methods, elaboration of theory of viral carcinogenesis, and the role of chronic inflammation facilitate the understanding of molecular carcinogenesis. The knowledge in the genetic and molecular interactions will aid in the establishment of accurate diagnosis and further prognostication in the era of personalized medicine.

Keywords

Oral squamous cell carcinoma · Oncogenes · Tumor suppressor genes · Transcription factor · Signaling pathway · Loss of heterogeneity

M. N. Sable (✉)
Department of Pathology, All India Institute of Medical Sciences, Bhubaneswar, Odisha, India
e-mail: pathol_mukund@aiimsbhubaneswar.edu.in

S. V. Kane
Department of Pathology, Tata Memorial Hospital, Mumbai, India
e-mail: drsvkane@rediffmail.com

S. Routray (ed.), *Microbes and Oral Squamous Cell Carcinoma*,
https://doi.org/10.1007/978-981-19-0592-6_3

3.1 Introduction

Oral squamous cell carcinoma (OSCC) is an aggressive malignancy with a shorter 5-year survival rate [1]. It is not a homogenous disease in etiology, molecular pathogenesis, clinical presentations, and behavior [2]. OSCC from different anatomical sites is usually a result of multiple molecular events developed in the lining squamous epithelium on exposure to environmental carcinogens, which are also influenced by an individual's genetic predisposition [3]. Chronic exposures to multiple carcinogens, i.e., tobacco, alcohol, oncogenic viruses, inflammation, and genetic predisposition of individuals, induce attrition in genes significantly responsible for subsequent molecular events, etiopathogenesis and disease progression. The variable and unpredictable response to various treatment modalities for OSCC at different sites is primarily because of molecular heterogeneity [4]. There is a paradigm shift in the understanding of molecular pathogenesis of OSCC with the biological perspective of the diseases, which gets additional translational value in targeted therapy and prognostication of OSCC [5, 6].

3.2 Mechanisms of Molecular Carcinogenesis of OSCC

Like other epithelial malignancies, OSCC development is a constellation of multiple processes in a stepwise manner. There is a sequential accumulation of genetic and epigenetic changes with changes in subsequent encoding of the proteins involved in signaling pathways which lead to cancer-associated phenotypes. These genetic alterations bring two significant processes, namely "Inactivation of tumor suppressor genes" and "Activation of oncogenes." Inactivation of tumor suppressor genes occurs by mutations, deletions, loss of heterozygosity, or epigenetic modifications such as chromatin remodeling and methylation of DNA. On the other hand, activation of oncogenes occurs mainly by overexpression due to increased transcription, gene amplification, or structural changes.

All the above-mentioned mechanisms have been established in the development of OSCC. The resultant genetic alterations affect cell cycle regulation, immunological responses, and growth factors which together lead to histologic progression from epithelial hyperplasia to dysplasia with further progression into invasive cancer and metastasis.

Thus, a working molecular model for the multistep process of carcinogenesis has been proposed [7–9]. The details of well-established oncogenes, tumor suppressor genes, cell cycle regulators, and growth factors are summarized in Table 3.1.

The multistep process of molecular carcinogenesis of OSCC results from direct interaction of mucosa with various etiologic agents to which it is exposed. The main carcinogenic agents are listed below.

Table 3.1 Oncogenes, tumor suppressor genes, cell cycle regulators, and growth factors involved in oral carcinogenesis

Sr No	Name	Changes	Type	Effects	Step in carcinogenesis
Tumor suppressor gene					
1	p53 [10, 11]	Mutation	Point mutation	Non-activation of p21 transcription	Initiator of carcinogenesis
			Deletion	Reduction or Loss of p53 expression	–
2	doc-1 [12]	LOH	–	Reduction of expression and protein function	Unclear
3	Retinoblastoma gene (Rb) [13]	LOH	–	–	Unclear, progression from CIS to SCC
4	VHL [14, 15], FHIT [16]	Loss of chromosome 3p	Loss-of-function mutations	Accumulation of HIF-1α	OSCC progression
			Promoter hypermethylation	–	Epithelial-mesenchymal transition
5	APC [17]	LOH with mutation	–	–	Unclear
Oncogenes					
a	Growth factor				
1	TGF-α [18, 19]	Aberrant expression	Overexpression, autocrine stimulation	Stimulates cell proliferation	Early carcinogenesis, hyperplasia to dysplasia
				Stimulates angiogenesis	Invasive carcinoma progression
2	TGF-β [20]	Aberrant expression	Overexpression	Defective TGF-β signaling	Tumor progression and metastasis
b	Cell surface receptors				
1	EGFR [21]	Aberrant expression	Deletion or mutations	Overexpression of EGFR	Early carcinogenesis
			Amplification	–	–

(continued)

Table 3.1 (continued)

Sr No	Name	Changes	Type	Effects	Step in carcinogenesis
c	Signal transduction pathways				
1	H-RAS [22]	Mutation	Point mutations	Activation of RAS	Proliferation and survival of cancer cells
				Activation of PI3K signaling pathway	–
2	BRAF [23, 24]	Mutation	Point mutation	Constitutive activation of MEK and ERK	Proliferation and survival of cancer cells
d	Proteins transactional activator				
1	C-MYC [25]		Amplification	Overexpression of CMYC	The proliferation of the cell, control in differentiation, and regulation of apoptosis
e	Cell cycle regulators				
	Cyclin D1 [26]	Aberrant expression	Amplification	Activates cell cycle progression	Late stages of carcinogenesis
			Overexpression		
f	Inhibitors of apoptosis				
	Bcl2 [27]	Abnormal Bcl-2 expression	Overexpression	Cell survival and inhibition of apoptosis	Early events in carcinogenesis, cell proliferation

3.3 Exposure to a Carcinogen

Chronic exposure to chemical and biological carcinogens such as tobacco, alcohol, oncogenic viruses, and unresolved inflammation initiates damage of individual genes and significant portions of chromosomes and subsequent attrition in genetic material.

3.3.1 Tobacco

Both chewing [smokeless] tobacco and tobacco smoking damage the genes of the lining epithelial cells directly or by generating intermediate metabolites. More than 16 known intermediate metabolites in smokeless tobacco and about 60 in smoke tobacco have been identified, which are toxic in nature and exhibit carcinogenic properties [28]. The mechanisms are multifocal.

The important carcinogens are tobacco-specific nitrosamines, polycyclic aromatic hydrocarbons, and aromatic amines. The potency and effect of these carcinogens are variable depending upon the type and form of tobacco and its method of consumption.

Tobacco acts as a stimulant for the generation of more free radicals, which are the by-product of many cellular metabolic processes involved in the responses to the bacteria.

3.3.1.1 Mechanism of Action

The free radicals involve reactive oxygen species (ROS), reactive nitrogen species (RNS), and reactive oxygen metabolites such as hydrogen peroxide (H_2O_2), superoxide anions (O_2^-), hydroxyl radicals (OH^-), nitric oxide (NO), and malondialdehyde. They can induce several DNA changes by causing damage to mucosa, including strand breakage, DNA-protein cross-linkage, and base modification [29]. ROS and other metabolites also generated changes in the oral microbiome, contributing to carcinogenesis [30].

Alteration in local immune response induced by tobacco metabolites and increase in susceptibility to Epstein-Barr infection also trigger early changes of carcinogenesis in OSCC [31, 32].

Apart from the effect of free radicals, tobacco also influences the overexpression of tumor suppressor genes like P53, P16, GLUT-1, DAPK, MGMT, and PI3K by inducing epigenetic modifications leading to progression [33].

3.3.2 HPV

The role of HPV infection in the oncogenesis of several epithelial tumors has been firmly established in the last century. HPV is a double-stranded DNA virus with an epitheliotropic property that mainly infects the stratified squamous cells of the skin and mucosal lining. In the oral cavity, HPV infection has predilection for the lining of tonsillar crypts and the stratified squamous epithelium, which is less susceptible to oncogenic transformation compared to other sites. However, its role in the pathogenesis of OSCC is now convincingly established with high-risk HPV subtypes (i.e., HPV16, HPV18, HPV31, HPV33, HPV35, HPV39, HPV45, HPV51, HPV52, HPV56, HPV58, HPV59, and HPV68) [34]. Both lineage and non-lineage-specific variation of E6 and L1 regulatory regions or coding sequence of the HPV16 are mainly associated with OSCC [35].

Viral DNA is incorporated into the host genome and initiates malignant transformation by E6 and E7 oncogenes. In infected epithelium, E6 and E7 oncogenes increase phosphorylation and degradation of p53 and RB gene, ultimately resulting in the escape of cell cycle checkpoints and further cellular proliferation. Genetic variant brings transforming function of these oncoproteins by inducing gene transcription or increasing amino acid exchange. Similarly, E6 amino acid variation changes the affinity to p53 protein by either increasing or decreasing tumor suppressor function.

In HPV-induced OSCC, other than p53 and pRb pathways (cell cycling), different vital pathways have also been implicated, which are associated with carcinogenesis as well as survival, i.e. (1) the EGFR pathway (growth factor signaling), (2) the TGF-β pathway (growth factor signaling), (3) PI3K–PTEN–AKT (evading apoptosis), and (4) angiogenesis including hypoxia-inducible factor (HIF) [36].

In HPV-induced OSCC, T350G is the most prevalent E6 genomic variant with a high viral load [37]. In addition to the viral transfection of the epithelium, epigenetic or genetic malfunction of RB gene involved in the p16INK4A–cyclin D1–RB pathway progresses to the development of OSCC [38].

3.3.3 Chronic Inflammation

The link between chronic inflammation and carcinoma was conceptualized nearly five decades ago. Viral pathogens like *Epstein-Barr virus (EBV), human papillomavirus (HPV)*, and fungi *Candida* are identified in precancerous lesions like leukoplakia. The pieces of evidence clearly suggest that the inflammatory response and battery of cytokines produced in the peritumoral tissue might contribute to carcinogenesis and progression.

The primary pathology in the progression of inflammatory **lesions** to carcinoma is the local cytokine expression profile, which induces immunosuppression and changes in transcription factors like nuclear factor-kB, r signal transducers, activators of transcription-3 (STAT-3), and activator protein-1 (AP-1). These activations in noncancerous tissue of inflammatory conditions may increase cell proliferation and cell survival, promoting initial cancerous transformation. The malignant transformation of OLP and OLK may thus be related to, or dependent on, a series of molecular events occurring within inflamed tissue. The infiltration of inflammatory cells also helps in generating ROS and reactive nitrogen species, causing DNA damages to bring initiation of tumorigenesis [39].

3.4 Progression of Premalignant Lesions and Malignant Transformation

The premalignant lesions in the oral cavity are conditions, which carry a definite risk of OSCC development. The most common premalignant lesion is the oral leukoplakia [OLP] and its variants (conventional leukoplakia and proliferative verrucous leukoplakia), erythroplakia, and submucous fibrosis. These precancerous lesions are associated with oral epithelial dysplasia (OED) and OSCC.

The hyperplasia-dysplasia-early-advanced-carcinoma sequence may provide a reference panel of markers for defining premalignant lesions and predicting the risk of malignant transformation and tumor progression (Fig. 3.1).

3.4.1 Alteration in Oncogenes

In oral leukoplakia, DNA damage repair gene defects (i.e., BRCA1, BRCA2, FANCA, DCLRE1B, NEIL3) are strongly associated with progressive lesions irrespective of dysplasia [40]. There is a substantial loss of Rb expression pattern from normal mucosa to hyperplasia with a significant increase in cyclin D1 and Ki67 [41]. Alterations in KMT2C and TP53 and inactivation of CDKN2A (P16INK4a/P14ARF), which code for the tumor suppressor proteins, are frequently involved in PVL which is the rare special type.

Transforming growth factor-α (TGF-α) is overexpressed early in oral carcinogenesis by hyperplastic squamous epithelium.

3.4.2 Alteration in Chromosomes

DNA aneuploidy is associated with a 1.32-fold increased risk for oral leukoplakia to cancer development. In addition to DNA aneuploidy, chromosomal instability has also been observed in progressive verrucous leukoplakia (PVL) to predict its risk of developing into carcinoma. Loss of heterogeneity (LOH) at 3p and 9p increases 3.8-to-8.5-fold risk of malignant transformation, which may further increase to 33-fold after additional LOH at 4q, 8p, 11q, 13q, and 17p.

In comparing hyperplastic epithelium with dysplasia, the most common genetic alteration in OSCC is loss of chromosomal region 9p21, representing CDKN2A locus encoding tumor suppressors p16 and p14ARF [3, 42]. Another common early genetic aberration in oral carcinogenesis is the loss of the chromosome 3p region, encoding FHIT and RSSFIA [43, 44]. Progressive accumulation of genetic aberrations and their specific pattern of occurrence is essential for advanced modifications in the sequence of oral carcinogenesis [3].

In the early stages of OSCC oncogenesis, dysregulation of MAPK, ERK, JNK, IL-6/STAT3, WNT, TGFβ, and glucocorticoid receptor signaling may initiate OSCC tumorigenesis in premalignant lesions. ILK, IGF1R, HIF1, MAPK, and WNT pathways are dysregulated to gather malignant transformations in leukoplakia

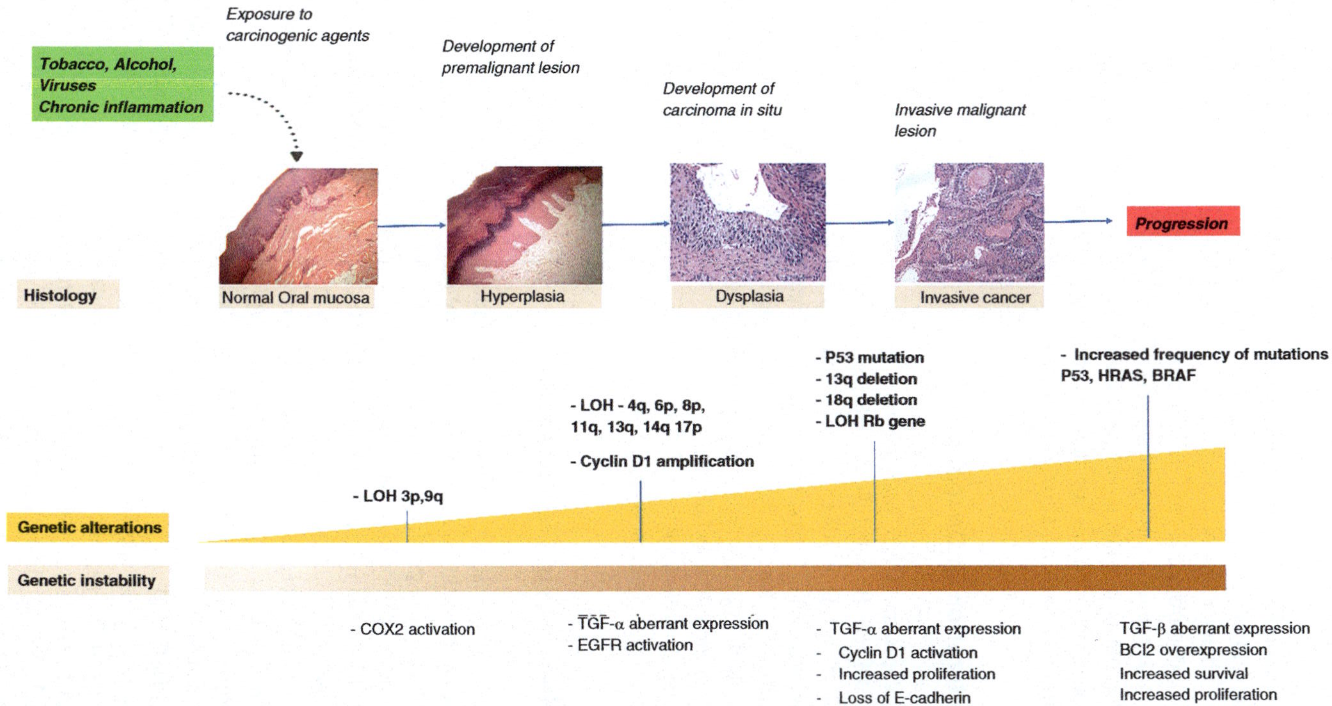

Fig. 3.1 Genetic and histological alterations in hyperplasia-dysplasia-early-advanced-carcinoma sequence of oral carcinogenesis

[45]. With further progression of dysplastic epithelium, the significant expression of EGFR, STAT3, c-MYC, p53, BCL2, H-ras, and loss of p16 become more evident leading to carcinoma.

These inappropriate and complexly deregulated expressions of the various oncoproteins contribute to the early stages of oral carcinogenesis. Amplification of 11q13 and increased expression of cyclin D1 have been noted in around 40% of dysplastic squamous epithelium [46].

In the later stage of carcinogenesis, the genetic alteration associated with progression from dysplasia to invasive OSCC is loss of heterozygosity (LOH) of chromosome 17p and mutation of p53. These alterations of p53 are mainly mutation or deletion. Further p53-induced genomic instability in dysplastic epithelium may accelerate the additional genetic modifications for further progression [47, 48].

Eventually, it was established that the increase in the genetic alterations (DNA aneuploidy, copy number variation, and LOH) is proportional to the severity of dysplastic changes and the risk of malignant transformation.

3.5 Invasive Carcinoma and Progression of the OSCC

The molecular mechanisms involved in the pathogenesis of OSCC progression are necessary to define the patient's prognosis. In the initial step, carcinoma cells invade the basement membrane and detach from the primary site to invade the underlying tissue further.

After dissociation of cell-cell adhesion and invasion, the phenotype of cancer cells and its interaction with the microenvironment play an essential role in the OSCC progression. At the invasive front, the cancer cells transform from epithelial to mesenchymal cell-like phenotype, referred to as epithelial-mesenchymal transition (EMT) [49].

Though the cadherin switch is not well established in OSCC, a decrease or loss of E-cadherin expression in the membrane of oral cancer cells from dysplasia to carcinoma suggests E-cadherin's role as a tumor suppressor gene. Activation of the PI3K/Akt signaling pathway and stimulation of EGFR were shown to be associated with induction of EMT in OSCC [50, 51]. EMT is characterized by downregulation of E-cadherin, β-catenin, desmoplakin and upregulation of vimentin, resulting in a reduction in cell-cell adhesion, increased cell motility on fibronectin-coated surfaces, and increased invasiveness properties [52].

The primary important tumor suppressor gene involved in OSCC progression is mutant p53, by facilitating the function of the key transcriptional regulators of EMT, TWIST1, and SLUG, which promotes EMT. The gain of function of p53 also enhances TGFβ-mediated metastasis in the late stage of progression.

3.6 Conclusion

The molecular carcinogenesis of OSCC is a heterogeneous, complex, and multistep process in terms of etiology, molecular alterations, and further involvement of interconnected signaling networks. Modifications in signaling networks occur in the form of overexpression, under expression of growth factors or cell surface receptors, proto-oncogene activation, transcriptional changes, or loss of tumor suppressor activity. These modifications are essential for the cellular processes of proliferation, differentiation, regulation of apoptosis, cell migration, epithelial and mesenchymal transition. These processes are the basic steps of the multistep carcinogenesis involving tumor initiation, progression, and metastasis of OSCC.

References

1. Kumar M, Nanavati R, Modi TG, Dobariya C (2016) Oral cancer: etiology and risk factors: a review. J Cancer Res Ther 12(2):458–463. https://doi.org/10.4103/0973-1482.186696. PMID: 27461593
2. Leemans CR, Braakhuis BJ, Brakenhoff RH (2011) The molecular biology of head and neck cancer. Nat Rev Cancer 11(1):9–22. https://doi.org/10.1038/nrc2982. PMID: 21160525
3. Califano J, van der Riet P, Westra W, Nawroz H, Clayman G, Piantadosi S, Corio R, Lee D, Greenberg B, Koch W, Sidransky D (1996) Genetic progression model for head and neck cancer: implications for field cancerization. Cancer Res 56(11):2488–2492. PMID: 8653682
4. Alsahafi E, Begg K, Amelio I, Raulf N, Lucarelli P, Sauter T, Tavassoli M (2019) Clinical update on head and neck cancer: molecular biology and ongoing challenges. Cell Death Discov 10(8):540. https://doi.org/10.1038/s41419-019-1769-9. PMID: 31308358; PMCID: PMC6629629
5. Cancer Genome Atlas Network (2015) Comprehensive genomic characterization of head and neck squamous cell carcinomas. Nature 517(7536):576–582. https://doi.org/10.1038/nature14129. PMID: 25631445; PMCID: PMC4311405
6. Walter V, Yin X, Wilkerson MD, Cabanski CR, Zhao N et al (2013) Molecular subtypes in head and neck cancer exhibit distinct patterns of chromosomal gain and loss of canonical cancer genes. PLoS One 8(2):e56823. https://doi.org/10.1371/journal.pone.0056823
7. Todd R, Donoff RB, Wong DT (1997) The molecular biology of oral carcinogenesis: toward a tumor progression model. J Oral Maxillofac Surg 55(6):613–623. https://doi.org/10.1016/s0278-2391(97)90495-x. discussion 623–625. PMID: 9191644
8. Williams HK (2000) Molecular pathogenesis of oral squamous carcinoma. Mol Pathol 53(4): 165–172. https://doi.org/10.1136/mp.53.4.165. PMID: 11040937; PMCID: PMC1186964
9. Krishna A, Singh S, Kumar V, Pal US (2015) Molecular concept in human oral cancer. Natl J Maxillofac Surg 6(1):9–15. https://doi.org/10.4103/0975-5950.168235. PMID: 26668446; PMCID: PMC4668742
10. Shin DM, Charuruks N, Lippman SM, Lee JJ, Ro JY, Hong WK, Hittelman WN (2001) p53 protein accumulation and genomic instability in head and neck multistep tumorigenesis. Cancer Epidemiol Biomark Prev 10(6):603–609. PMID: 11401909
11. Cruz IB, Snijders PJ, Meijer CJ, Braakhuis BJ, Snow GB, Walboomers JM, van der Waal I (1998) p53 expression above the basal cell layer in oral mucosa is an early event of malignant transformation and has predictive value for developing oral squamous cell carcinoma. J Pathol 184(4):360–368. https://doi.org/10.1002/(SICI)1096-9896(199804)184:4<360:AID-PATH1263>3.0.CO;2-H. PMID: 9664901
12. Winter J, Pantelis A, Reich R, Jepsen S, Allam JP, Novak N, Wenghoefer M (2011) Risk estimation for a malignant transformation of oral lesions by S100A7 and Doc-1 gene

expression. Cancer Investig 29(7):478–484. https://doi.org/10.3109/07357907.2011.597813. PMID: 21740085
13. Soni S, Kaur J, Kumar A, Chakravarti N, Mathur M, Bahadur S et al (2005) Alterations of rb pathway components are frequent events in patients with oral epithelial dysplasia and predict clinical outcome in patients with squamous cell carcinoma. Oncology 68(4–6):314–325. https://doi.org/10.1159/000086970. PMID: 16020958
14. Zhang S, Zhou X, Wang B, Zhang K, Liu S, Yue K et al (2014) Loss of VHL expression contributes to epithelial-mesenchymal transition in oral squamous cell carcinoma. Oral Oncol 50(9):809–817. https://doi.org/10.1016/j.oraloncology.2014.06.007. PMID: 24998140
15. Scully C, Field JK, Tanzawa H (2000) Genetic aberrations in oral or head and neck squamous cell carcinoma 3: clinico-pathological applications. Oral Oncol 36(5):404–413. https://doi.org/10.1016/s1368-8375(00)00023-3. PMID: 10964046
16. Tanimoto K, Hayashi S, Tsuchiya E et al (2000) Abnormalities of the *FHIT* gene in human oral carcinogenesis. Br J Cancer 82:838–843. https://doi.org/10.1054/bjoc.1999.1009
17. Tsuchiya R, Yamamoto G, Nagoshi Y, Aida T, Irie T, Tachikawa T (2004) Expression of adenomatous polyposis coli (APC) in tumorigenesis of human oral squamous cell carcinoma. Oral Oncol 40(9):932–940. https://doi.org/10.1016/j.oraloncology.2004.04.011. PMID: 15380172
18. Wong DT (1993) TGF-alpha and oral carcinogenesis. Eur J Cancer B Oral Oncol 29B(1):3–7. https://doi.org/10.1016/0964-1955(93)90003-w. PMID: 8180574
19. Srinivasan M, Jewell SD (2001) Evaluation of TGF-alpha and EGFR expression in oral leukoplakia and oral submucous fibrosis by quantitative immunohistochemistry. Oncology 61(4):284–292. https://doi.org/10.1159/000055335. PMID: 11721175
20. Pang X, Tang YL, Liang XH (2018) Transforming growth factor-β signaling in head and neck squamous cell carcinoma: insights into cellular responses. Oncol Lett 16(4):4799–4806. https://doi.org/10.3892/ol.2018.9319. PMID: 30250544; PMCID: PMC6144927
21. Scully C (1993) Oncogenes, tumor suppressors and viruses in oral squamous carcinoma. J Oral Pathol Med 22(8):337–347. https://doi.org/10.1111/j.1600-0714.1993.tb01086.x. PMID: 8283399
22. Saranath D, Chang SE, Bhoite LT, Panchal RG, Kerr IB, Mehta AR, Johnson NW, Deo MG (1991) High frequency mutation in codons 12 and 61 of H-ras oncogene in chewing tobacco-related human oral carcinoma in India. Br J Cancer 63(4):573–578. https://doi.org/10.1038/bjc.1991.133. PMID: 2021541
23. Bruckman KC, Schönleben F, Qiu W, Woo VL, Su GH (2010) Mutational analyses of the BRAF, KRAS, and PIK3CA genes in oral squamous cell carcinoma. Oral Surg Oral Med Oral Pathol Oral Radiol Endod 110(5):632–637. https://doi.org/10.1016/j.tripleo.2010.05.002. PMID: 20813562; PMCID: PMC2987741
24. Weber A, Langhanki L, Sommerer F, Markwarth A, Wittekind C, Tannapfel A (2003) Mutations of the BRAF gene in squamous cell carcinoma of the head and neck. Oncogene 22(30):4757–4759. https://doi.org/10.1038/sj.onc.1206705. PMID: 12879021
25. Mastronikolis N, Ragos V, Kyrodimos E, Chrysovergis A, Papanikolaou V, Mastronikolis S, Stamatelopoulos A, Tsiambas E (2019) Mechanisms of C-myc oncogenic activity in head and neck squamous cell carcinoma. J BUON 24(6):2242–2244. PMID: 31983089
26. Ramos-García P, Gil-Montoya JA, Scully C, Ayén A, González-Ruiz L, Navarro-Triviño FJ, González-Moles MA (2017) An update on the implications of cyclin D1 in oral carcinogenesis. Oral Dis 23(7):897–912. https://doi.org/10.1111/odi.12620. PMID: 27973759
27. Kang MH, Reynolds CP (2009) Bcl-2 inhibitors: targeting mitochondrial apoptotic pathways in cancer therapy. Clin Cancer Res 15(4):1126–1132. https://doi.org/10.1158/1078-0432.CCR-08-0144. PMID: 19228717; PMCID: PMC3182268
28. Hecht SS (2003) Tobacco carcinogens, their biomarkers and tobacco-induced cancer. Nat Rev Cancer 3(10):733–744. https://doi.org/10.1038/nrc1190. Erratum in: Nat Rev Cancer. 2004;4(1):84. PMID: 14570033

29. Korde SD, Basak A, Chaudhary M, Goyal M, Vagga A (2011) Enhanced nitrosative and oxidative stress with decreased total antioxidant capacity in patients with oral precancer and oral squamous cell carcinoma. Oncology 80(5–6):382–389. https://doi.org/10.1159/000329811. PMID: 21829039
30. Lissoni A, Agliardi E, Peri A, Marchioni R, Abati S (2020) Oral microbiome, and mucosal trauma as risk factors for oral cancer: beyond alcohol and tobacco. A literature reviews. J Biol Regul Homeost Agents 34(6 Suppl. 3):11–18. PMID: 33386052
31. Gaur P, Mittal M, Mohanti B, Das S (2011) Functional variants of IL4 and IL6 genes and risk of tobacco-related oral carcinoma in high-risk Asian Indians. Oral Dis 17(7):720–726. https://doi.org/10.1111/j.1601-0825.2011.01831.x. PMID: 21771210
32. Nasher AT, Al-Hebshi NN, Al-Moayad EE, Suleiman AM (2014) Viral infection, and oral habits as risk factors for oral squamous cell carcinoma in Yemen: a case-control study. Oral Surg Oral Med Oral Pathol Oral Radiol 118(5):566–572.e1. https://doi.org/10.1016/j.oooo.2014.08.005. PMID: 25442493
33. Jiang X, Wu J, Wang J, Huang R (2019) Tobacco, and oral squamous cell carcinoma: a review of carcinogenic pathways. Tob Induc Dis 17:29. https://doi.org/10.18332/tid/105844. PMID: 31582940; PMCID: PMC6752112
34. Syrjänen S, Lodi G, von Bültzingslöwen I, Aliko A, Arduino P, Campisi G, Challacombe S, Ficarra G, Flaitz C, Zhou HM, Maeda H, Miller C, Jontell M (2011) Human papillomaviruses in oral carcinoma and oral potentially malignant disorders: a systematic review. Oral Dis 17(Suppl 1):58–72. https://doi.org/10.1111/j.1601-0825.2011.01792.x. PMID: 21382139
35. Mirabello L, Clarke MA, Nelson CW, Dean M, Wentzensen N, Yeager M, Cullen M, Boland JF, NCI HPV Workshop, Schiffman M, Burk RD (2018) The intersection of HPV epidemiology, genomics and mechanistic studies of HPV-mediated carcinogenesis. Viruses 10(2):80. https://doi.org/10.3390/v10020080. PMID: 29438321; PMCID: PMC5850387
36. Rautava J, Syrjänen S (2012) Biology of human papillomavirus infections in head and neck carcinogenesis. Head Neck Pathol 6(Suppl 1):S3–S15. https://doi.org/10.1007/s12105-012-0367-2. PMID: 22782219; PMCID: PMC3394166
37. Hoffmann M, Lohrey C, Hunziker A, Kahn T, Schwarz E (2004) Human papillomavirus type 16 E6 and E7 genotypes in head-and-neck carcinomas. Oral Oncol 40(5):520–524. https://doi.org/10.1016/j.oraloncology.2003.10.011. PMID: 15006625
38. Buenahora MR, Lafaurie GI, Perdomo SJ (2021) Identification of HPV16-p16^{INK4a} mediated methylation in oral potentially malignant disorder. Epigenetics 16(9):1016–1030. https://doi.org/10.1080/15592294.2020.1834923. PMID: 33164635; PMCID: PMC8451454
39. Feller L, Altini M, Lemmer J (2013) Inflammation in the context of oral cancer. Oral Oncol 49(9):887–892. https://doi.org/10.1016/j.oraloncology.2013.07.003. PMID: 23910564
40. Farah CS, Jessri M, Bennett NC, Dalley AJ, Shearston KD, Fox SA (2019) Exome sequencing of oral leukoplakia and oral squamous cell carcinoma implicates DNA damage repair gene defects in malignant transformation. Oral Oncol 96:42–50. https://doi.org/10.1016/j.oraloncology.2019.07.005. PMID: 31422212
41. Shah NG, Trivedi TI, Tankshali RA, Goswami JA, Shah JS, Jetly DH, Kobawala TP, Patel KC, Shukla SN, Shah PM, Verma RJ (2007) Molecular alterations in oral carcinogenesis: significant risk predictors in malignant transformation and tumor progression. Int J Biol Markers 22(2):132–143. PMID: 17549669
42. van der Riet P, Nawroz H, Hruban RH, Corio R, Tokino K, Koch W, Sidransky D (1994) Frequent loss of chromosome 9p21-22 early in head and neck cancer progression. Cancer Res 54(5):1156–1158. PMID: 8118798
43. Kisielewski AE, Xiao GH, Liu SC, Klein-Szanto AJ, Novara M, Sina J, Bleicher K, Yeung RS, Goodrow TL (1998) Analysis of the FHIT gene and its product in squamous cell carcinomas of the head and neck. Oncogene 17(1):83–91. https://doi.org/10.1038/sj.onc.1201910. PMID: 9671317

44. Dong SM, Sun DI, Benoit NE, Kuzmin I, Lerman MI, Sidransky D (2003) Epigenetic inactivation of RASSF1A in head and neck cancer. Clin Cancer Res 9(10 Pt 1):3635–3640. PMID: 14506151
45. Makarev E, Schubert AD, Kanherkar RR, London N, Teka M, Ozerov I, Lezhnina K, Bedi A, Ravi R, Mehra R, Hoque MO, Sloma I, Gaykalova DA, Csoka AB, Sidransky D, Zhavoronkov A, Izumchenko E (2017) *In silico* analysis of pathways activation landscape in oral squamous cell carcinoma and oral leukoplakia. Cell Death Discov 3:17022. https://doi.org/10.1038/cddiscovery.2017.22. PMID: 28580171; PMCID: PMC5439156
46. Rousseau A, Lim MS, Lin Z, Jordan RC (2001) Frequent cyclin D1 gene amplification and protein overexpression in oral epithelial dysplasias. Oral Oncol 37(3):268–275. https://doi.org/10.1016/s1368-8375(00)00114-7. PMID: 11287281
47. Brennan JA, Boyle JO, Koch WM, Goodman SN, Hruban RH, Eby YJ, Couch MJ, Forastiere AA, Sidransky D (1995) Association between cigarette smoking and mutation of the p53 gene in squamous-cell carcinoma of the head and neck. N Engl J Med 332(11):712–717. https://doi.org/10.1056/NEJM199503163321104. PMID: 7854378
48. Saleem S, Abbasi ZA, Hameed A, Qureshi NR, Khan MA, Azhar A (2014) Novel p53 codon 240 Ser > Thr coding region mutation in the patients of oral squamous cell carcinoma (OSCC). Tumour Biol 35(8):7945–7950. https://doi.org/10.1007/s13277-014-2062-2. PMID: 24833091
49. Kalluri R, Weinberg RA (2009) The basics of epithelial-mesenchymal transition. J Clin Invest 119(6):1420–1428. https://doi.org/10.1172/JCI39104. Erratum in: J Clin Invest. 2010;120(5):1786. PMID: 19487818; PMCID: PMC2689101
50. Zuo JH, Zhu W, Li MY, Li XH, Yi H, Zeng GQ, Wan XX, He QY, Li JH, Qu JQ, Chen Y, Xiao ZQ (2011) Activation of EGFR promotes squamous carcinoma SCC10A cell migration and invasion via inducing EMT-like phenotype change and MMP-9-mediated degradation of E-cadherin. J Cell Biochem 112(9):2508–2517. https://doi.org/10.1002/jcb.23175. PMID: 21557297
51. Grille SJ, Bellacosa A, Upson J, Klein-Szanto AJ, van Roy F, Lee-Kwon W, Donowitz M, Tsichlis PN, Larue L (2003) The protein kinase Akt induces epithelial mesenchymal transition and promotes enhanced motility and invasiveness of squamous cell carcinoma lines. Cancer Res 63(9):2172–2178. PMID: 12727836
52. Krisanaprakornkit S, Iamaroon A (2012) Epithelial-mesenchymal transition in oral squamous cell carcinoma. ISRN Oncol 2012:681469. https://doi.org/10.5402/2012/681469. PMID: 22548191; PMCID: PMC3324906

Hallmarks of Cancer: Its Concept and Critique

4

Niharika Swain, Rashmi Hosalkar, Mansee Thakur, and Antony Herold Prabhu

Abstract

The complicated chronobiology of carcinogenesis has been organized by concept of cancer hallmarks. Though this concept has already mapped out the established/emerging hallmarks and enabling characteristics contributing to multistep genesis of human tumors with greater detail, still the clarity on signaling circuit describing the intercell communication in tumors needs to be answered along with the need of justification on the critique over its universal clinical implication in tumor management. Apart from these controversies, cancer hallmarks have been successfully wired all the traits and characteristics contributing to tumorigenesis. Among these characteristics, tumor promoting inflammation seems to play decisive role in all stages of tumor development. It also plays an intermediate link between biologic carcinogens (infections) with human cancer. This book chapter attempted to provide an overview of cancer hallmark and principal mechanisms elaborating the role of inflammation and infections in cancer biology.

N. Swain (✉)
(Medical Genetics, MGM School of Biomedical Sciences, MGMIHS), Department of Oral Pathology, MGM Dental College and Hospital, Navi Mumbai, Maharashtra, India
e-mail: niharikadec30@gmail.com

R. Hosalkar
Department of Oral Pathology, MGM Dental College and Hospital, Navi Mumbai, Maharashtra, India
e-mail: drrashmi009@gmail.com

M. Thakur
Department of Medical Biotechnology, MGM School of Biomedical Sciences, MGMIHS, Navi Mumbai, India
e-mail: mansibiotech79@gmail.com

A. H. Prabhu
H Lee Moffitt Cancer Center, Tampa, FL, USA
e-mail: antonyherold@gmail.com

S. Routray (ed.), *Microbes and Oral Squamous Cell Carcinoma*,
https://doi.org/10.1007/978-981-19-0592-6_4

Keywords

Cancer · Infection · Inflammation · tumorigenesis

4.1 Cancer Hallmark: Its Concept and Critique

Nearly two decades ago, Douglas Hanahan and Robert A Weinberg attempted to conceptualize the acquired capabilities and characteristics of cancer as its hallmarks known as Hallmark I. For the first time, they proposed six major hallmarks demonstrating multistep carcinogenesis like (1) sustaining the proliferative signals, (2) evading growth suppressors, (3) resisting cell death, (4) enabling replicative immortality, (5) inducing angiogenesis, and (6) activating invasion and metastasis. This revolutionary review publication laid down a solid foundation for understanding the complexity and diversities of cancer biology [1]. A decade later, owing to remarkable developments in cancer research and growing importance of mechanistic concepts, they amended the Hallmark I to Hallmark II by adding two new emerging hallmark capabilities and two enabling characteristics/traits, supposed to be critical for acquiring the six core hallmark capabilities. De-regulating cellular energetics and avoiding immune destruction have been adopted only as emerging hallmarks. Although both reprogrammed energy metabolism and immune evasion demonstrated promising influence on cancer biology, these could not be added to core hallmark characteristics as altered energy metabolism is functionally dependent on core hallmarks and research on antitumor immunity still in elementary state. As the hallmarks of cancer denote as acquired functional capabilities by cancer cells which allow them to survive, proliferate, and disseminate through different mechanisms during the multistep carcinogenesis, in Hallmark II, the authors introduced two enabling characteristics which thought to help the cancer cells acquiring these capabilities. First one is genome instability and mutation which describes the development of mutations and chromosomal rearrangements generated by genetic instability in cancer cells and their potency to orchestrate the hallmark capabilities. The second enabling characteristic is tumor promoting inflammation which states the capability of inflammatory state driven by host innate immunity in progression of premalignant and malignant lesions. In addition to these enabling characteristics and emerging hallmarks, another interesting and promising concept of "Tumor microenvironment" (TME) has been introduced in hallmark II. Interaction of mesenchymal cells which forms tumor-associated stroma with neoplastic epithelial compartment through regulatory signaling pathways is highlighted in the recent review—Hallmark of cancer: The next generation [2].

In past years, many researchers have made critic reviews on concept of cancer hallmarks which claims cancer to be a cell-based genetic disease, characterized by uncontrolled cell proliferation and governed by somatic mutation theory (SMT) [3]. Few authors questioned its goal of providing organizational framework of cellular properties as it failed to uncover the cause of the cellular alteration and also the details of malignant transformation process responsible for developing a premalignant or malignant lesion. Some have advocated for tissue organization field theory in which cancer is thought to be a tissue-based disease and carcinogenesis is believed because of alterations in paracrine effect among cells and interactions between tumor cells and the extracellular matrix. Universal applicability of proposed hallmarks of cancer is always under scanner as complexity of cancer is further complicated by variation in cancer risk among different tissues affecting tumor progression rate. Some authors have also attempted to organize the hallmarks by their own understanding through modifying or adding some characteristics as selective growth and proliferative advantage, altered stress response favoring overall survival, metabolic rewiring, abetting microenvironment, and immune modulation whereas retaining two of them, i.e., vascularization and invasion and metastasis [4]. With new developments in cancer research mainly focused on tumor management, the clinical success of concept of cancer hallmarks is going to be tested over time and more revision on this concept could be anticipated. Till then, the H & W (Hanahan & Weinberg) theory could be regarded as a dominant strategy for war against cancer in terms of understanding the mechanism of carcinogenesis and planning for effective tumor management protocol. In this chapter, we have tried to draw a connecting link between various infective pathogens and carcinogenesis through intermediary role of inflammation as an enabling characteristics of cancer hallmark (Fig. 4.1).

4.2 Role of Inflammation

4.2.1 Inflammation in Cancer Hallmark: An Overview

Inflammation is a physiologic protective mechanism targeted towards the removal of internal and external stimuli and maintenance of tissue homeostasis, mainly driven by vascular system of an organism. It includes a series of events like stagewise recruitment of immune cells and secretions of functional molecules under the influence of regulatory pathways leading to strengthening the host defense against infection with unwanted outcome as tissue damage. In 1863, the notorious linkage between cancer and inflammation has gained attention as Virchow first proposed about the association of lympho-reticular infiltration with origin of carcinogenesis. Following this revelation, research work targeting the intersection between inflammation and tumorigenesis has amplified through years to demonstrate the compelling evidence of tumor promoting effects of inflammation. In 2011, inflammation has been termed as an enabling characteristic of cancer hallmarks as it can potentially contribute in attaining multiple hallmark capabilities by delivering various bioactive molecules like growth factors, survival factors, proangiogenic, extracellular matrix modifying agents, and inductive signals [2].

Various clinical and experimental studies have already illustrated the association of chronic inflammatory conditions of microbial and nonmicrobial origin with human carcinomas [5]. The microenvironment created by chronic inflammation which tosses between continuous tissue damage and cellular proliferation proves to be deleterious as it is capable of producing mutagenic agents. Some of these

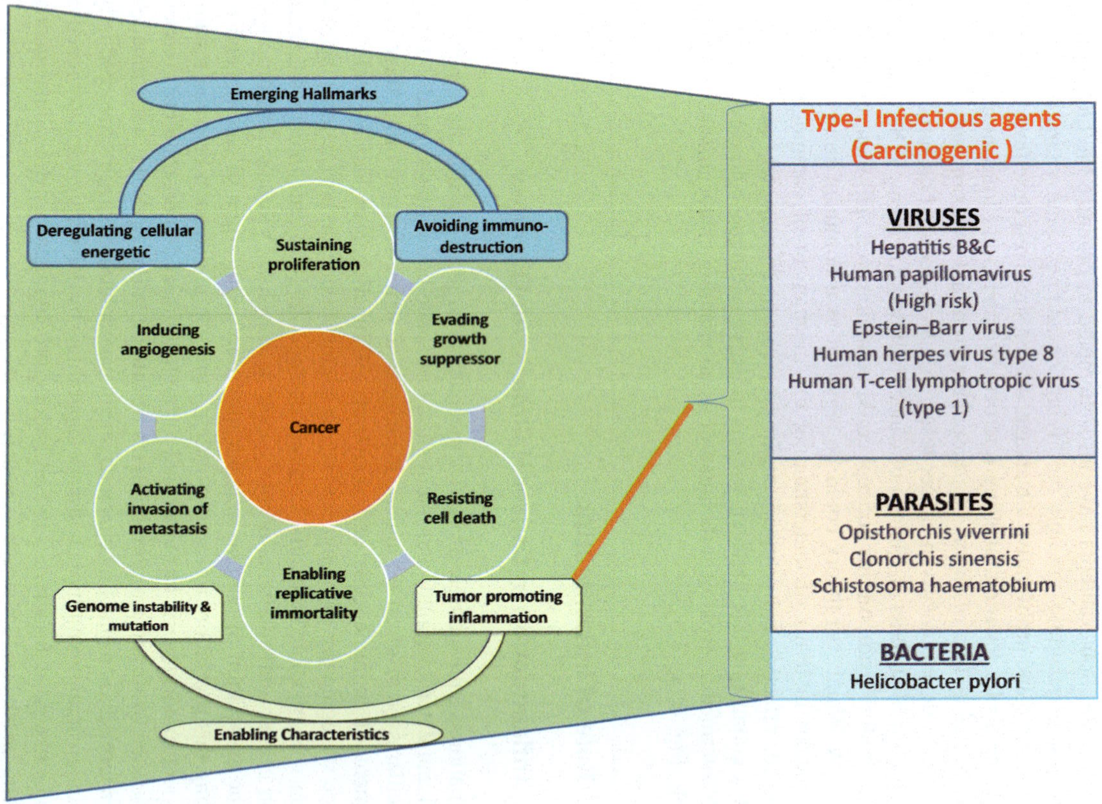

Fig. 4.1 Tumor promoting inflammation as a crucial link between various carcinogenic infectious agents and cancer hallmark, though various traits and characteristics of cancer hallmark could be altered by these agents with or without its contribution

agents like pro-inflammatory cytokines mainly interleukin-1, tumor necrosis factor α, and macrophage migration inhibitory factors can induce oncogenic mutation in native epithelial and stromal cells, thus contributing to cancer initiation. Apart from genetic mutation, epigenetic events like angiogenesis in association with chronic inflammation play a key role in tumor progression. Inflammatory TME can also contribute in invasion and metastasis of tumor cells by facilitating basement membrane breakdown in epithelial tumors [6].

4.2.2 Key Players of Tumor Promoting Inflammation

4.2.2.1 Immune Cells

Exact role of immune cells in cancers is a topic of debate since their association have surfaced. Several hypotheses have been put forward advocating their pro- or antitumoral effect. One of the promising hypotheses explains contribution of cells of innate and adaptive immunity in tumorigenesis through various steps like selection of aggressive clones, induction of immunosuppression and stimulation of cancer cell proliferation and finally metastasis. In initial stages of tumorigenesis, the cytotoxic immune cells like CD8+ T cells, macrophages, neutrophils, and natural killer cells eliminate the immunogenic cancer cells leaving behind proliferation of special clone of cells which are less immunogenic and therefore unrecognizable by immune system. In later stage, other inflammatory cells take up their specific role deciding the biological course of the tumor.

4.2.2.1.1 Neutrophils

In physiological state, neutrophils are first immune cells to be recruited to the site of injury and capable of modulating the process of inflammation by several mechanisms like phagocytosis, release of neutrophil extracellular traps (NETs) along with secretion of antibacterial proteins and protease containing granules. In cancer, neutrophils have gained the attention as association of neutrophilia, high neutrophil to lymphocyte ratio, and increased tumor associated neutrophils (TANs) with poor prognosis of various malignancies. TANs also have been proposed in two polarized forms N1 and N2 with putative protumoral and antitumoral properties, respectively. In cancer initiation, TANs contribute to cancer initiation through release of cytokine IL17 and elastase which in turn promotes cell proliferation and invasion. By secretion of MMP-9 and VEGF, TANs mediate angiogenic drive in TME. Recently, cancer promoting role of NETs which is an extracellular network released by neutrophils is also highlighted by researchers. NETs composed of chromatin, proteases, and intracellular proteins are associated with progression of metastasis by trapping the circulating cancer cells and hence favoring their extravasation and colonization.

4.2.2.1.2 Macrophages

Tumor associated macrophages (TAMs) are thought to be involved in various stages of carcinogenesis from cancer initiation to metastatic progression like neutrophils,

TAMs also exist in two polorized forms like pro-inflammatory M1 and anti-inflammatory M2. In initial stages of carcinogenesis, TAMs exhibit M1 like polarization which involves in elimination of immunogenic cancer cells. On progression of tumor, a change of polarization to M2 type elicits tumor promoting role of TAMs through a wide range of effects like cell proliferation, extracellular matrix modeling, epithelial mesenchymal transition, and immunosuppressive action on other antitumor immune cells [5].

4.2.2.1.3 Natural Killer Cell

NK cells are known to be a potent innate immune cell, being key players of immune surveillance. Notable antitumoral activity of natural killer cells has been appreciated in initial stages of cancer transformation. When activated, these granulated cells mediate tumor suppressing activity by releasing cytotoxic perforin and granzyme helpful in elimination of tumor cells. They also trigger apoptic pathways in tumor cells through the production of TNF-α or via direct cell–cell contact through activation of the TRAIL and FASL pathways.

4.2.2.1.4 T cells

T cells found to be second most frequent immune cells in TME after TAM as extensive studies on potential role of T cells in various steps of carcinogenesis have been carried out in last decades. In cancer initiation, T cells have a prominent anitumor activity as they eliminate immunogenic cancer cells in TME. Upon activation by antigen presenting cells, CD8+ T cells, a prominent cytotoxic T cells carry out tumor cell killing through exocytosis of perforin and granzyme containing granules. On the other hand, CD4+ T cells have a wide range of antitumoral activity as it not only accelerates the priming and activation of T cells pro-inflammatory cytokines such as IL-2, TNF-α, and IFN-γ, but also modulates the tumor cell elimination by macrophages and NK cells. As the tumor grows, vital changes occur in TME. One of such changes is recruitment of a new subset of T cells, i.e., regulatory CD4+ T cells (Tregs) which are responsible for suppressing the antitumor activity of majority of effector immune cells by blocking their priming, activation, and cytotoxicity.

4.2.2.1.5 Dendritic Cells

These specialized cells help in presenting the antigens to T cells and thus connect innate and adaptive immunity. Though the exact role of dendritic cells (DCs) in tumorogenesis is understudied, their indirect contribution has been well documented as they can modulate or trigger antitumor T cell response.

4.2.3 Mediators and Molecular Pathways in Tumor Promoting Inflammation

Inflammation acts as a key regulator of tumor promotion and progression by several mechanisms including acceleration of cell cycle progression and cell proliferation, evasion from apoptotic cell death, and stimulation of tumor neovascularization. Many inflammatory mediators and cellular pathways have been identified in inflammation-cancer linkage. Among the major molecular players involved in the inflammation-to-cancer axis, the prominent members are cytokines, cyclooxygenase-2 (Cox-2), inducible nitric oxide synthase (iNOS), and NF-kB.

4.2.3.1 Mediators

In the transition of chronic inflamed tissue to premalignancy, many regulatory molecules or mediators play their specific roles.

4.2.3.1.1 Cytokines

Cytokines, membrane bound molecules play key role in differentiation and activation of immune cells. They can be broadly classified as pro-inflammatory (IL-1, IL-6, IL-17) and anti-inflammatory cytokines (IL-10). Pro-inflammatory contribution of cytokines can be explained by two mechanisms: induction of cell proliferation and inhibition of apoptosis in altered cells at inflammatory site. IL-6 is one of major inflammatory cytokine thought to play a substantial role in inflammation-related carcinogenesis. IL-6 seems to modulate the gene expression which is involved in cell cycle progression and inhibition of apoptosis, through the JAK-STAT signaling pathway [7]. In addition, IL1, another potent pro-inflammatory cytokine can contribute in carcinogenesis through the activation of master transcription factor like NF-kB and hence promotes the expression of genes involved in cell survival, proliferation, and angiogenesis. IL-17 also has proangiogenic properties. Chemokines are soluble chemotactic cytokines, expressed by both infiltrated leukocytes and tumor cells in an autocrine or a paracrine fashion in inflamed TME. Out of major groups of chemokines, CXC and CC act on directional migration of lymphocytes. Neutrophils are targeted only by CXC chemokines, whereas CC chemokines can also act on other immune cells such as monocytes, eosinophils, dendritic and NK cells. Besides their indirect role in chemoattraction of leukocytes, chemokines also directly attribute in metastasis through induction of matrix metalloproteinases (MMPs) and collagenases and facilitating tumor invasion. Tumor necrosis factor (TNF), another potent cytokine is secreted mainly by activated macrophages and by tumor cells. In carcinogenesis, TNF can induce DNA damage, inhibition of DNA repair, induction of antiapoptotic pathways, promotion of angiogenesis and finally provide growth factors for tumor cells.

4.2.3.1.2 Cyclooxygenase-2 (Cox-2)

This inducible form of cyclooxygenase enzyme has been largely studied in the context of its link with oncogenesis. In response to external stimuli like lipopolysaccharides, reactive oxygen species, UV rays, and inflammatory cytokines, alteration of tissue level of Cox-2 can induce various oncogenic functions like cell proliferation, inhibition of apoptosis, induction of angiogenesis, and invasion. Cox-2 enzyme mediates arachidonic acid metabolism which produces series of pro-inflammatory prostaglandins (PGE_2, PGF_2, and 15d-PGJ_2), which are found to be key mediators of inflammation-cancer axis. Among all, PGE2 directly and indirectly promotes cell proliferation, angiogenesis, and inhibits apoptosis through various cellular pathways thereby favoring tumor progression.

4.2.3.1.3 Inducible Nitric Oxide Synthase (iNOS)

This enzyme is another important linker between chronic inflammation and cancer as it synthesizes NO endogenously in macrophages and epithelial cells through arginine metabolism. NO and its derivative peroxynitrite play vital tumor promoting roles in inflammation-associated carcinogenesis by inducing damage to DNA, post-translational modification of p53, suppression of DNA repair enzymes like human thymine-DNA glycosylase, inhibition of apoptosis by targeting caspases, activation of HIF-1α-VEGF mediated angiogenesis and metastasis.

4.2.3.1.4 NF-kB

Mounting evidence of constitutive association with pro-inflammatory cytokines and other inflammatory mediators suggest NF-kB, dimeric transcription factors of the Rel family as critical mediator in inflammation-induced carcinogenesis. These factors contribute to genetic instability through induction of mutation by promoting ROS production. In addition, aberrant activation of NF-kB contributes to tumorigenesis by transactivating several target genes like immune-mediating genes and inflammatory genes, antiapoptotic genes, and cell proliferation regulation genes. NF-kB-mediated activation of adhesion molecules, such as E-selectin, vascular cell adhesion molecule-1, and intercellular adhesion molecule-1 induces leukocyte adhesion and migration, which are important in both inflammation and the inflammatory microenvironment of cancer and facilitate tumor cell migration and localization in the process of metastasis.

4.2.3.2 Molecular Pathways

Two prominent pathways, i.e., intrinsic (oncogene-driven) and extrinsic (microenvironment-driven), have been described in context with positive association of inflammatory reaction with neoplastic progression of altered tissue. In the intrinsic pathway, oncogene activation mediates the development of cancers through an inflammatory microenvironment, whereas in the extrinsic pathway, inflammatory conditions in microenvironment promote oncogenesis. Transcription factors (NF-kB, HIF-1α), immune cells and soluble mediators (cytokines, chemokines, and Cox-2) act as common effectors for both pathways to link inflammation with cancer.

The intrinsic pathway hints about connection between occurrence of genetic events and inflammation. The genetic alteration like activation of oncogenes and inactivation of tumor suppressor genes may trigger the pro-inflammatory cascade. Examples of activated oncogenes-inflammatory mediator pairs found to be correlated in tumor progression through eliciting inflammation like epidermal growth factor receptor (EGFR)-Cox2, ras oncogene-CXCL8, myc-mast cell. Similarly, inactivation of tumor suppressor proteins like von Hippel Lindau/hypoxia-inducible factor (VHL/HIF), transforming growth factor-β (TGF-β), and phosphatase and tensin homolog (PTEN) induces tumor-associated inflammation in different cancers through various mediators, i.e., TNF-α, CXCR4, CXCL5/CXCL12, respectively. In these ways, genetic alterations contribute to the cancer-inflammation linkage process.

In external pathway, TME-invaded leukocytes in inflammatory conditions have been held responsible for producing mediators. A wide range of inflammatory conditions like chronic infections, autoimmune diseases, and exposure to noxious stimuli have been positively associated with carcinogenesis. For example, inflammatory bowel disease and chronic airway inflammatory conditions (asbestosis, silicosis, pulmonary fibrosis, tuberculosis, etc.) which are linked with increased risk of colorectal cancer and non-tobacco-associated lung cancer, respectively. Besides inflammatory conditions, ROS, nitrogen mediators, and mitochondrial products released by damaged cells can activate inflammatory cytokines in addition to DNA damage, thus considered to be upstream regulators in inflammation-cancer related signaling pathways.

4.3 Role of Infections

First official recognition of controversial link between cancer and infectious agents came up with an interesting discovery by Nobel prize winner Johannes Fibiger in 1926 as he observed causative association of a nematode parasite with stomach cancer. Since then, many research works have attempted to explore the interactions between pathogens and cancer [8]. In 2018, approximately 2.2 million new cases of infection-attributable cancers have been documented worldwide, which was 13% of all cancer cases [9]. Among diseases with infective etiology, cancers are related to chronic infections with mainly virus, whereas bacteria and parasites also contribute significantly.

4.3.1 Viral Infections and Cancer

Various human viruses, namely human herpes virus (HHV), human papillomavirus (HPV), hepatitis B virus (HBV), hepatitis C virus (HCV), and human T-cell lymphotrophic virus (HTLV) have been consistently linked to various cancers, which attributes to 10–15% of global cancer burden. In 1958, Dennis Burkitt suggested the viral etiology for pediatric B cell malignancy. Later, Anthony Epstein and Yvonne Barr identified a new type of herpes virus linked to B cell lymphoma as Human Herpes Virus-4 (HHV-4) or Epstein–Barr virus (EBV). Apart from B cell lymphoma, EBV infection is related with different human malignancies, including African Burkitt's lymphoma, Hodgkin's lymphoma, nasopharyngeal carcinoma, and also benign disease like infectious mononucleosis mainly affecting the Asian population. In B cell malignancies, though exact causative role of EBV virus remains unclear, it is suggested that EBV infection causes B lymphocyte proliferation by immortalizing them. Latent membrane protein 1 (LMP-1), transmembrane oncoprotein of EBV appears to mediate this B cell population expansion. Cytoplasmic carboxyl terminus of LMP-1 binds with TNF receptor associated factor and TNF receptor associated death domain protein, which in turn activate key signaling molecules JNK, PI3K, JAKs, and transcription factors like NF-kB, AP-1, and STATs responsible for cell proliferation. A new addition to this HHV family came as HHV-8 was positively associated with Kaposi's sarcoma and other lymphoproliferative diseases like primary effusion lymphoma and multicentric Castleman's disease. HHV-8 is otherwise known as KSHV (Kaposi's sarcoma associated herpes virus). The genome of KSHV is comprised of wide number of cellular genes, i.e., cell cycle control, antiapoptotic, immune system regulation and cellular signaling or communication which are responsible for its proangiogenic and oncogenic potential. Hepatitis B virus (HBV) has a causative role in hepatocellular carcinoma which is one of the leading causes of deaths in every year. One of HBV encoded proteins, X protein, thought to activate molecular mechanisms involved in proliferation and inflammation including activation of JAK1, protein kinase C, PI3K, and MAPK and directly interfere with EGR1, OCT1, RXR, and CREB. Similarly, HCV can cause sustained liver damage and inflammation through immunological reaction which ultimately leads to liver cirrhosis and cancer [10]. One of the possible mechanisms for HCV-induced carcinogenesis may be the compromised repair of DNA damage in altered cells rendering them more susceptible for mutagenic alterations. Over the decades, many studies revealed relationship between heterogeneity in the HPV and their association with various benign and malignant neoplasms. HPVs are classified into low- or high-risk groups depending upon the likelihood of inducing cervical cancer on persistent infection. Majority of cervical cancers are linked with high-risk HPV-16 and 18 through expression of viral oncogenes like E6 and E7. E6 and E6-AP induce the rapid degradation of p53 through the proteasome pathway on ubiquitination, whereas HPV-encoded E7 alone can lead to deregulated cellular proliferation by inactivation of pRb. This step can be counteracted by p53-induced apoptosis. But the concomitant expression of E6 proteolytic degradation of p53 via formation of a trimeric E6/E6AP/p53 complex blocks apoptotic response and results in abnormal cellular growth [11, 12]. Another oncogenic virus, Human T-cell leukemia virus (HTLV-1) has been regarded as causative agent for adult T-cell leukemia. Unlike other oncogenic retroviruses, HTLV-1 neither carry a host-derived oncogene nor activate cellular proto-oncogenes through insertional mutation. The virus starts the oncogenic process by infecting CD4 lymphocytes through its transactivator protein Tax. Plethora of effects of Tax involves (1) expression of several cytokines IL-2 and IL-15 and (2) causing alteration in cell cycle (3) inhibition of apoptosis and (4) promotes genetic instability. It also inhibits DNA repair which causes aneuploidy [13].

4.3.2 Bacterial Infection and Cancer

Since eighteenth century, bacterial pathogens have been linked with cancers. After the first unsuccessful attempt to prove the cause-and-effect relationship between tuberculosis and lung cancer, various other cancer models have been identified in this context. Though the direct attribute of organisms to cancer has not uncovered fully, the nonspecific mechanisms explaining the putative role of bacteria in carcinogenesis have been much explored. The first mechanism is induction of inflammation which is best explained by association of Helicobacter pylori infection and gastric adenocarcinoma. *H. pylori* was declared as Group-1 carcinogen by the International Agency for Research on Cancer (IARC) and contributes to carcinogenesis with age standardized incidence rates of 8.7 cancer cases per 100,000 person-years [9]. Oncogenic potential of this bacterium can be explained by its ability to induce chronic inflammation and thus indirectly influencing the cell proliferation rate and causing alteration in terminal cell differentiation. Damage to gastric epithelium can be direct (through release of ammonia or by inducing gastrin production) and indirect (through inflammatory response due to ROS and cytokine IL-8 production). Increased cell proliferation in turn induces DNA replication error and thus predisposes mucosal cells to malignant transformation through dietary or endogenous mutagens [14]. Other neoplasms like B cell MALT lymphoma have also been associated with H. Pylori infection. Second mechanism linking the pathogen to cancer is explained by production of mutagenic compounds though bacterial metabolism as depicted in colon cancer model. Several theories and observations have been put forward to explain this mechanism like bacterial mediated deconjugation of 7a-hydroxyl groups from bile acids to produce cytotoxic 7ax-dehydroxylating bile acids and hence inducing cell proliferation and promoting the growth of adenomas. Bacteria are also thought to activate the exogenous procarcinogens like rutin to quercetin (a mutagenic aromatic amine) and hydrolysis of cycasin to methylazoxymethanol [15].

4.3.3 Parasitic Infections and Cancer

Parasitic infections caused by helminths and protozoa are considered to be one of the predominant health problems mainly in developing countries. Being infectious agents, parasites have been positively associated with carcinogenesis due to their potential to induce tumor inflammation in host through free radical induced DNA damage. Few examples for such well-established tumorigenic association of parasites are Schistosoma haematobium-bladder cancer, Opisthorchis viverrini and Clonorchis sinensis-cholangiocarcinoma, Trichomonas vaginalis-cervical cancer, and Cryptosporidium parvum-gastrointestinal cancers [16]. However, their antitumor capabilities have been highlighted by few authors.

4.3.3.1 Pro-tumorigenic Role of Parasites

Out of 11 species of Group I biological (IARC) carcinogens, three have been identified as parasitic agents, i.e., Opisthorchis viverrini, Clonorchis sinensis, and Schistosoma haematobium [17]. Although there is no direct cause and effect relationship established between other species of Opisthorchis and Schistosoma and different parasitic agents, i.e., malaria and Trypanosoma cruzi with cancer, their potential causative roles have been postulated by several investigators time and again [18, 19].

Various mechanisms of action have been explained by authors to elaborate the pro-tumorigenic role of parasites. Other than the well-established carcinogenic mechanism of chronic inflammatory infection process, parasitic infections are thought to be involved in additional alteration in genetic and proteomic level to pursue their tumorigenic potential [16]. It has been postulated about the release of pro-inflammatory and carcinogenic metabolites, i.e., oxysterols and catechol estrogen by helminths like Schistosoma haematobium and Opisthorchis viverrini. These metabolites can form DNA adduct through their complex interaction with host DNA [20]. Apart from their role in initiation stage of carcinogenesis, parasitic infections are found to modulate the antitumor immuno-surveillance response and thus contribute to clonal expansion of malignant tumor cells and inhibition of tumor rejection [21]. It is an interesting note that parasitic infections can accelerate the process carcinogenesis by influencing nearly all cancer hallmarks, emerging hallmarks, and enabling characteristics [22] (Fig. 4.2).

4.3.3.2 Anti-Tumorigenic Role of Parasites

Among meticulous description of tumor promoting role played by some parasitic diseases, cumulating evidence from several in vitro and in vivo studies highlighted antitumorigenic potential of the parasitic infections in contrast. Though the mechanism of action explaining the anticarcinogenic effect of parasites still needs to be scrutinized and validated, their regulatory effects of carcinogenesis cannot be ignored. Direct interference in growth and proliferation of transformed cells and indirect contribution in modulation of cancer developments remain the key observations in anticarcinogenic role of parasites. Unspecific activation of immune cells, induction of cytokines, reduction of chemokines, and inhibitory effect on cell cycle is some of the elucidated antitumor mechanisms of parasitic infections helping in cancer elimination. It has been also claimed that antitumor mechanisms of these helminths and protozoa target almost half of cancer hallmarks [23]. Despite of these promising inhibitory roles in carcinogenesis, need of more detailed studies is advocated to explore the effects of the parasites and their products on intracellular signaling pathways, so that the future use of these potent biological factors as adjuvants to current anticancer drugs could be another possibility to combat cancer.

4.4 Conclusion

A sequential link could be drawn between infections and carcinogenesis through inflammation as several infectious agents have been found to be involved in various stages of carcinogenesis through tumor promoting inflammation which is an enabling characteristic of cancer hallmark. As cancer is complex and multistep process, key research on regulatory effects of all the contributory agents/molecules of carcinogenesis should continue exploring their precise antitumor/protumor

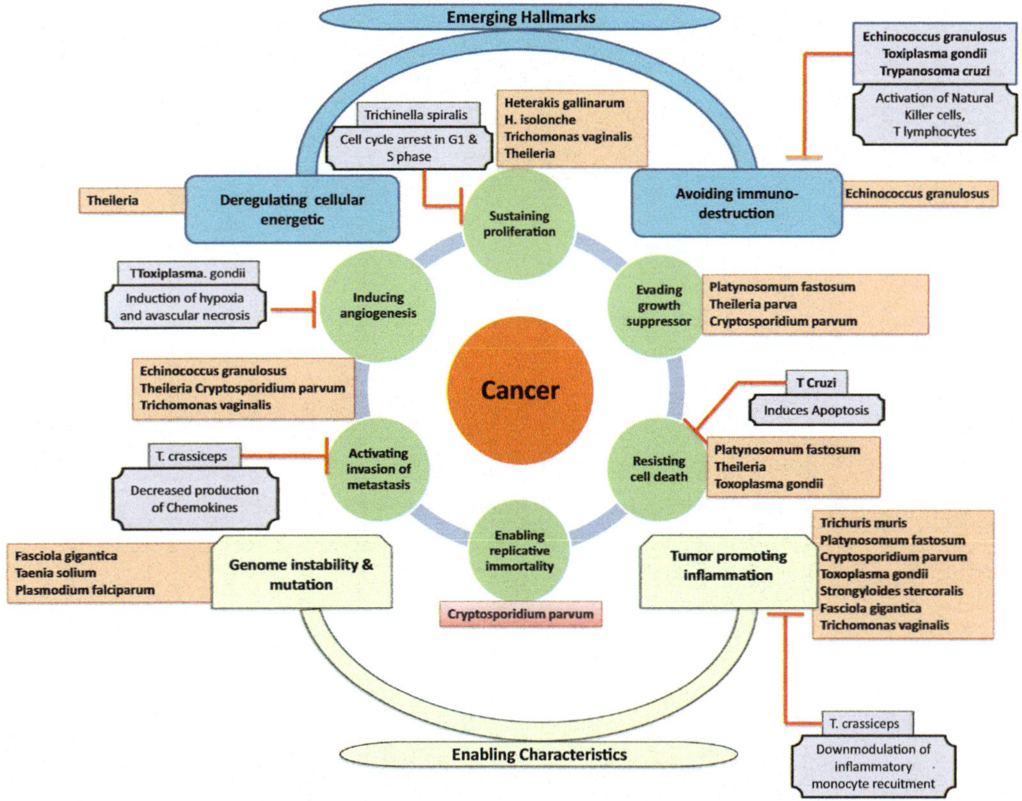

Fig. 4.2 A detailed representation of pro-tumorigenic and anti-tumorigenic role of various parasites affecting the hallmarks and enabling characteristics of cancer

mechanism, so that clinical implication of the observations can be utilized in prevention, diagnosis, treatment, and prognosis of human cancer.

References

1. Hanahan D, Weinberg RA (2000) Hallmarks of cancer. Cell 100:57–70
2. Hanahan D, Weinberg RA (2011) Hallmarks of cancer : the next generation. Cell 144(5): 646–674. https://doi.org/10.1016/j.cell.2011.02.013
3. Sonnenschein C, Soto AM (2014) The aging of the 2000 and 2011 Hallmarks of cancer reviews: a critique. J Biosci 38(3):651–663
4. Fouad YA, Aanei C (2017) Revisiting the hallmarks of cancer. Am J Cancer Res 7(5): 1016–1036
5. Allavena P, Garlanda C, Borrello MG, Sica A, Mantovani A (2008) Pathways connecting inflammation and cancer. Curr Opin Genet Dev 18:3–10
6. Lu H, Ouyang W, Huang C, Lu H, Ouyang W, Huang C (2006) Inflammation, a key event in cancer development. Mol Cancer Res 4(4):221–233
7. Kumar J, Surh Y (2008) Inflammation: gearing the journey to cancer. Mutat Res 659:15–30
8. Darveau R (1999) Infection, inflammation, and cancer. Nat Biotechnol 17:1999
9. De Martel C, Georges D, Bray F, Ferlay J, Clifford GM (2020) Global burden of cancer attributable to infections in 2018 : a worldwide incidence analysis. Lancet Glob Health 8(2): e180–e190. https://doi.org/10.1016/S2214-109X(19)30488-7
10. Jesuino L, Andrade DO, Junior ADO, Melo RC, De Souza EC, Alves C et al (2009) Association between hepatitis C and hepatocellular carcinoma. J Global Infect Dis 1(1):33–37
11. Martin D, Gutkind JS (2009) Human tumor-associated viruses and new insights into the molecular mechanisms of cancer. Oncogene 27:31–42
12. Hoppe-seyler K, Bossler F, Braun JA, Herrmann AL, Hoppe-seyler F (2017) The HPV E6/E7 oncogenes: key factors for viral carcinogenesis and therapeutic targets. Trends Microbiol 26(2): 158–168. https://doi.org/10.1016/j.tim.2017.07.007
13. Mahieux R, Gessain A (2007) Adult T-cell leukemia/lymphoma and HTLV-1. Curr Hematol Malig Rep 2:257–264
14. Parsonnet J (1995) Bacterial infection as a cause of cancer. Environ Health Perspect 103(8): 263–268
15. Mackowiak P (1982) The normal microbial flora. N Engl J Med 307(2):83–93
16. Machicado C, Marcos LA (2016) Carcinogenesis associated with parasites other than schistosoma, opisthorchis and clonorchis: a systematic review. Int J Cancer 138:291521
17. IARC (2012) Biological agents. A review of human carcinogens. IARC Monogr Eval Carcinog Risks Hum 100(Pt B):1–441
18. Krementsov N (2009) *Trypanosoma cruzi*, cancer and the cold war. Hist Cienc Saude Manguinhos 16(Suppl. 1):75–94
19. Molyneux EM, Rochford R, Griffin B, Newton R, Jackson G, Menon G (2012) Burkitt's lymphoma. Lancet 379(9822):1234–1244
20. Gouveia MJ, Brindley PJ, Rinaldi G et al (2019) Infection with carcinogenic helminth parasites and its production of metabolites induces the formation of DNA-adducts. Infect Agents Cancer 14:41. https://doi.org/10.1186/s13027-019-0257-2
21. Vennervald BJ, Polman K (2009) Helminths and malignancy. Parasite Immunol 31:686–696. https://doi.org/10.1111/j.1365-3024.2009.01163.x
22. van Tong H, Brindley PJ, Meyer CG, Velavan TP (2017) Parasite infection, carcinogenesis and human malignancy. EBioMedicine 15:12–23. https://doi.org/10.1016/j.ebiom.2016.11.034
23. Callejas BE, Martínez-Saucedo D, Terrazas LI (2018) Parasites as negative regulators of cancer. Biosci Rep 38(5):BSR20180935. https://doi.org/10.1042/BSR20180935

Role of Inflammation and Infection in Oral Squamous Cell Carcinoma

5

Niharika Swain, Jay Gopal Ray, Rashmi Hosalkar, Mansee Thakur, and Kapil Singh Thakur

Abstract

Putative etiological association of various microbial infections with oral squamous cell carcinoma is gaining attention in research work. Various bacteria, viruses, and fungi have been claimed to oncogenic influence on oral mucosa mainly through dysregulated inflmmatory pathways. Chronic polymicrobial insult to oral mucosa along with altered immune status may be a possible explanation for its vulnerability to develop both oral potentially malignant disorders and oral carcinoma. Additionally, susceptibility of oral mucosa to physical, mechanical, and chemical trauma also plays an important role in

N. Swain (✉)
(Medical Genetics, MGM School of Biomedical Sciences, MGMIHS), Department of Oral Pathology, MGM Dental College and Hospital, Navi Mumbai, Maharashtra, India
e-mail: niharikadec30@gmail.com

J. G. Ray
Department of Oral and Maxillofacial Pathology, Burdwan Dental College and Hospital, Burdwan, West Bengal, India
e-mail: jaygopalray@yahoo.co.in

R. Hosalkar
Department of Oral Pathology, MGM Dental College and Hospital, Navi Mumbai, Maharashtra, India
e-mail: drrashmi009@gmail.com

M. Thakur
Department of Medical Biotechnology, MGM School of Biomedical Sciences, MGMIHS, Navi Mumbai, India
e-mail: mansibiotech79@gmail.com

K. S. Thakur
Datta Meghe Institute of Medical Sciences, Wardha, Maharashtra, India
e-mail: ksthakur123@gmail.com

S. Routray (ed.), *Microbes and Oral Squamous Cell Carcinoma*,
https://doi.org/10.1007/978-981-19-0592-6_5

disturbing its symbiotic relationship with oral microbiota. Though many organisms have been claimed to have oral oncogenic potential, the comprehensive research to elaborate their exact molecular mechanism is needed to access diagnostic, preventive, and therapeutic utility.

Keywords

Inflammation · Infection · Oral cancer · Microbes · Oral mucosa

5.1 Introduction

A sequential link has been already established between infections and carcinogenesis of various human organs through chronic inflammation. The persistent dysregulated inflammatory process incited by biological agents (virus, bacteria, and parasites) has the capacity to induce genetic instability via active mediators like cytokines, chemokines, prostaglandins, and reactive oxygen/nitrogen species. Once established, the inflammatory microenvironment not only promotes cancer initiation, but also contributes to transformed cell proliferation and tumor progression. Mounting evidence of positive correlation between chronic inflammation and malignant transformation risk in oral potentially malignant disorders also advocates its crucial role in oral carcinogenesis. Like other human cancers, the causative association of inflammation in initiation, promotion, and progression of oral squamous cell carcinoma (OSCC) has been highlighted in the literature [1]. In contrast, many authors have questioned the putative role of chronic inflammation as an exclusive cause in oral carcinogenesis as incidence of oral cancer does not seem to match with frequent prevalence of inflammatory conditions in oral cavity [2, 3]. This chapter aimed for outlining the role of inflammation and infection in pathogenesis of OSCC.

5.2 Inflammation and OSCC: A Close but Unsettled Relationship

Oral cavity is a special organ in the context of its ability of maneuvering the chronic irritation by several physical, chemical, and biological insults which are capable of induction inflammation. A variety of common extrinsic factors like cheek bite, irregular dental restorations, faulty brushing technique, plaque deposition, and ill-fitting dentures can induce inflammation. Persistent inflammation causes reshuffling the epithelial differentiation and keratinization through alteration in mucosal permeability [4]. In addition to these agents, deleterious tobacco habits with or without its catastrophic combination of alcohol also thought to use similar mechanism to trigger initial carcinogenic events in oral mucosa. Mucosal changes including development of premalignant lesions also witnessed a close association with the altered levels of proinflammatory mediators. An imbalance in immunohomeostasis is thought to play a key role in progression of mucosal inflammatory disease to oral cancer [5].

5.2.1 Role of Inflammation in Oral Cancer Initiation, Promotion, and Progression

Chronic inflammation induces oxidative damage in cells through production of reactive oxidative/nitrogen species generated by tobacco and alcohol use. Imbalance between oxidative stress and host anti-oxidative mechanism leads to production of mutagenic DNA adduct by the action of inflammation-driven cyclooxygenase and lipoxygenase enzymes mediated by arachidonic acid pathway. Inflammatory microenvironment is a constant source of oncogenic cytokines and mediators. Persistent presence of inflammatory mediators like IL-1β, IL6, IL-8, TNF-α, and COX2 in both oral potential malignant disorders and OSCC indicates about their putative role in initiation of oral carcinogenesis [6, 7].

Once oral cancer is established, the malignant oral keratinocytes and tumor-associated macrophages are thought to be source of proinflammatory cytokines like IL-1α, IL-1β, IL-6, and IL-8. Endogenous expression of IL-1α in OSCC seems to up-regulate the genes of IL-6 and IL-8 through autocrine activation of transcription factors like NF-kβ and activator protein (AP-1). The tumor cell-derived IL1α induces the proliferation of cancer-associated fibroblasts which in turn upregulates the expression of IL-6 and 8 [8]. COX-2 being an inflammation-induced enzyme is likely to be involved with major events of cancer promotion like induction of cancer stem cell like properties, promotion of angiogenesis, cell proliferation, inhibition apoptosis, and cell migration via maintenance of a chronic inflammatory state [9]. Another important mediator, TGF-β also seems to have major contribution in tumor promotion by engaging inflammation, angiogenesis, immunosuppression, and cell motility (epithelial mesenchymal transition development) [10].

Overexpression of the proinflammatory molecules in established oral cancer has been positively associated with poor prognosis, i.e., altered expression of IL-6 and IL-8 are indicative of invasive mode growth of OSCC. IL-8 acts as autocrine growth factor in head and neck squamous cell carcinoma (HNSCC) and is associated with oral cancer progression through promoting cell proliferation and angiogenesis [11]. Similarly, IL-6 overexpression in HNSCC patients with poor prognosis could be attributable to its contribution in tumor microenvironment immunosuppression via stimulation of myeloid-derived suppressor cells and PDL-1 expression [12]. Another oncogenic proinflammatory mediator, IL-1α has been positively correlated with increased incidence of metastasis in HNSCC as it has the capability to facilitate the transmigration of tumor cells through vascular endothelium and induce the expression of several potent metastatic genes like MMP-9, PGE2, and vascular endothelial growth factor (VEGF). Likewise, another member of IL-1 family, i.e., IL-1β, has been putatively associated with increase in tumor size and lymph node metastasis in OSCC. Similarly, COX2 in combination with VEGF has been identified as negative predictor of OSCC due to consistent correlation with lymph angiogenesis, lymph node metastasis, and TNM stage of OSCC [1, 13].

Increasing evidence suggesting causative association of inflammatory mediators (mainly NF-kβ dependent) in oral carcinogenesis, also attracted many researchers to explore their use as potential diagnostic and prognostic biomarkers in oral carcinogenesis. In this context, though supportive but diverging observations have been surfaced specifically in terms of their specificity and sensitivity to mark the progression of OPMDs to oral cancer. The suboptimal results are posing hindrance to their exclusive use as single diagnostic biomarker, rather future use of a panel of salivary biomarkers has been advocated [1].

5.3 Role of Infections in OSCC

The diverse and symbiotic microbiome plays a vital role in maintenance of oral health. Apart from common inflammatory conditions like gingivitis, periodontitis, or mucositis, linkup between oral microbiota and oral cancer has been reported by many authors. Imbalance in symbiosis of microbial community (includes commensal and opportunistic bacteria, fungi, and viruses) and host immune status has a major contribution in oral carcinogenesis resulted by chronic polymicrobial insult. Though the tobacco and alcohol use have causative association with oral carcinogenesis, these risk factors could not be able to explain the incidence of oral cancer in individuals without exposure of the same. So, in addition to molecular deregulation caused by chemical carcinogens, role of infectious agents in development and progression of oral cancer should be explored in depth.

5.3.1 Viruses and OSCC

Since the beginning of twentieth century, a consistent effort observed to explore the association between viruses and oral cancer through animal experiments. In 1985, first report about the impact of oncogenic virus, i.e., human papilloma virus (HPV), in oral and oropharyngeal cancers surfaced. Early HPV oncoproteins E6 and E7 are responsible for degradation of p53 and inactivation of retinoblastoma gene, thus causing cell cycle derangement and eventually to malignant transformation. Even if HPV has etiologically linked with virtually all cervical cancers, its association with oral cancer needs more clarification because of conflicting evidence [14]. Despite of the controversies with HPV positive OSCC, few contributing factors like gender (male), poor oral hygiene, and immunosuppression have been identified for increased risk of OSCC development in presence of HIV infection [15, 16]. Likewise, the concept of etiopathology association of Epstein–Barr Virus (EBV) with OSCC is underexplored because of inconsistent and inclusive observations reflecting differences in ethnicity, geography, and methodological sensitivity [17, 18].

5.3.2 Bacteria and OSCC

Several saccharolytic and acid tolerant bacteria like *Exiguobacterium oxidotolerans, Prevotella melaninogenica, Staphylococcus aureus*, and *Veillonella parvula* have been proposed to induce oral carcinogenesis. Mechanism of actions like direct and indirect interference in cell cycle, carcinogen metabolism, or induction of chronic inflammation have been explained in the context of bacteria-induced oral carcinogenesis. Owing to significant salivary specificity of bacteria like *Capnocytophagagingivalis, P. melaninogenica*, and *Streptococcus mitis*, their use as biomarker in OSCC development and prognosis has been advocated [19]. In addition, consistent attributes of *P. gingivalis* and *F. nucleatum has been* observed in oral cancer development and progression [20]. In spite of encouraging observations, an in-depth exploration of bacteria–OSCC relationship is mandatory to establish the etio-pathogenesis.

5.3.3 Fungi and OSCC

Direct causal role of fungi specially Candida albicans in OSCC development is not yet well established though its association has been reported time and again. Its ability to initiate oral cancer has been explained by its putative role in production of nitrosamines, a potent carcinogen. Binding of nitrosamine with DNA results in formation of DNA adduct and point mutation and leads to OSCC development. Strains of Candida with tubular hyphae morphology can produce N-nitrosobenzylmethylamine and are associated with initiation of oral cancer. Apart from this, an aetiopathologic link has been suggested between *C. albicans*, TLR and NF-κB, and finally results in the production of cytokines and COX-2, which can act as another potential mechanism explaining the association [21].

5.4 Conclusion

Carcinogenic potential of many oral micro-organisms is still under debate as oral cavity harbors many pathogenic microbiomes without any pathology. Though observations of many studies could not explain the exact causative role of these infectious agents in oral cancer development and progression, their consistent and increased presence in oral cancer tissues as compared with normal and benign counterpart supports their putative role in oral carcinogenesis. Well-structured studies focused on cause-and-effect relationship between infectious agents and oral cancer should be carried out to define their true oncogenic potential.

References

1. Niklander SE (2021) Inflammatory mediators in oral cancer: pathogenic mechanisms and diagnostic potential. Front Oral Health 2:1–10
2. Mantovani A (2010) Molecular pathways linking inflammation and cancer. Curr Mol Med 10(4):369–373
3. Grivennikov SI, Greten FR, Karin M (2010) Immunity, inflammation, and cancer. Cell 140(6): 883–899

4. Squier CA, Wertz PW (1993) Permeability, and the pathophysiology of oral mucosa. Adv Drug Deliv Rev 12(1–2):13–24
5. Sun Y, Liu N, Guan X, Wu H, Sun Z, Zeng H (2016) Immunosupression induced by chronic inflammation and the progression to oral squamous cell carcinoma. Mediat Inflamm 2016:1–12
6. Rao SK, Pavicevic Z, Du Z, Kim JG, Fan M, Jiao Y et al (2010) Pro-inflammatory genes as biomarkers and therapeutic targets in oral squamous cell carcinoma. J Biol Chem 285:32512–32521
7. Amirchaghmaghi M, Mohtasham N, Mozaffari PM (2012) Comparison of COX2 expression between oral squamous cell carcinoma, leukoplakia, and normal mucosa. J Contemp Dent Pract 13:205–209
8. Bates AM, Gomez Hernandez MP, Lanzel EA, Qian F, Brogden KA (2018) Matrix metalloproteinase (MMP) and immunosuppressive biomarker profiles of seven head and neck squamous cell carcinoma (HNSCC) cell lines. Transl Cancer Res 7:533–542
9. Nasry WHS, Rodriguez-Lecompte JC, Martin CK (2018) Role of COX-2/PGE2 mediated inflammation in oral squamous cell carcinoma. Cancers 10:348
10. Wu F, Weigel KJ, Zhou H, Wang XJ (2018) Paradoxical roles of TGF-β signaling in suppressing and promoting squamous cell carcinoma. Acta Biochim Biophys Sin 50:98–105. https://doi.org/10.1093/abbs/gmx127
11. Waugh DJ, Wilson C (2008) The interleukin-8 pathway in cancer. Clin Cancer Res 14:6735–6741
12. Tsai MS, Chen WC, Lu CH, Chen MF (2019) The prognosis of head and neck squamous cell carcinoma related to immunosuppressive tumor microenvironment regulated by IL-6 signaling. Oral Oncol 91:47–55
13. Morita Y, Morita N, Hata K, Nakanishi M, Kimoto N, Omata T et al (2014) Cyclooxygenase-2 expression is associated with vascular endothelial growth factor-c and lymph node metastasis in human oral tongue cancer. Oral Surg Oral Med Oral Pathol Oral Radiol 117:502–510
14. Hillbertz NS, Hirsch JM, Jalouli J, Jalouli MM, Sand L (2012) Viral and molecular aspects of oral cancer. Anticancer Res 32(10):4201–4212
15. Bui TC, Markham CM, Ross MW, Mullen PD (2013) Examining the association between oral health and oral HPV infection. Cancer Prev Res (Phila) 6:917–924
16. Mannarini L, Kratochvil V, Calabrese L, Gomes Silva L, Morbini P, Betka J et al (2009) Human papilloma virus (HPV) in head and neck region: review of literature. Acta Otorhinolaryngol Ital 29:119–126
17. Jalouli J, Ibrahim SO, Mehrotra R, Jalouli MM, Sapkota D, Larsson PA et al (2010) Prevalence of viral (HPV, EBV, HSV) infections in oral submucous fibrosis and oral cancer from India. Acta Otolaryngol 130:1306–1311
18. Jalouli J, Ibrahim SO, Sapkota D, Jalouli MM, Vasstrand EN, Hirsch JM et al (2010) Presence of human papilloma virus, herpes simplex virus and Epstein-Barr virus DNA in oral biopsies from Sudanese patients with regard totoombak use. J Oral Pathol Med 39:599–604
19. Chocolatewala N, Chaturvedi P, Desale R (2010) The role of bacteria in oral cancer. Indian J Med Paediatr Oncol 31(4):126–131. https://doi.org/10.4103/0971-5851.76195
20. Whitmore SE, Lamont RJ (2014) Oral bacteria and cancer. PLoS Pathog 10(3):e1003933
21. Mohd Bakri M, MohdHussaini H, Rachel Holmes A, David Cannon R, Mary Rich A (2010) Revisiting the association between candidal infection and carcinoma, particularly oral squamous cell carcinoma. J Oral Microbiol 2. https://doi.org/10.3402/jom.v2i0.5780

Normal Microbiome of the Oral Cavity

6

Supriya Kheur, Aneesha Acharya, Mohit Kheur, Vinay Jain, and Vinay K. Hazarey

Abstract

The microbiome residing in the complex niches throughout the oral cavity consists of a highly diverse microbiota of bacteria, fungi, and viruses. The interactions between the microbiome and human health are responsible for its innate and acquired immunity. This chapter provides an overview of the various aspects of oral microbiome like its acquisition, development, and maturation with a brief note on the various methods of identification and association.

S. Kheur (✉)
Department of Oral Pathology and Microbiology, D.Y. Patil Dental College and Hospital, Dr. D.Y. Patil Vidyapeeth, Pune, Maharashtra, India
e-mail: supriya.kheur@dpu.edu.in

A. Acharya
Department of Periodontics, Dr. D.Y. Patil Dental College and Hospital, Dr. D.Y. Patil Vidyapeeth, Pune, Maharashtra, India
e-mail: aneesha.acharya@dpu.edu.in

M. Kheur
Department of Prosthodontics, Rangoonwala College of Dental Sciences and Research Centre, Pune, Maharashtra, India
e-mail: mkheur@gmail.com

V. Jain
Department of Prosthodontics, University of Tennessee Health Sciences Centre College of Dentistry, Memphis, TN, USA
e-mail: vjain@uthsc.edu

V. K. Hazarey
Adjunct Faculty, Dr. D.Y. Patil Dental College and Hospital,Dr. D.Y.Patil Vidyapeeth, Pune, Maharashtra, India
e-mail: vinay.hazarey@gmail.com

S. Routray (ed.), *Microbes and Oral Squamous Cell Carcinoma*,
https://doi.org/10.1007/978-981-19-0592-6_6

Keywords

Oral microbiome · Niche · Acquisition · Oral cavity

6.1 Introduction

The oral cavity is the initial part of the digestive system with a very polymorphic and unique environment. Different niches from within the oral cavity have varied microbiome compositions. Various unique ecological niches are present in the oral cavity. As per evolutionary genetics, the oral cavity microflora is characterized into three forms: (1) Bacteria, (2) Archaea, (3) Eucarya [1]. The term microbiome is used because bacteria make up more than 99% of the total microbial flora, but the remaining less common flora constitutes a "rare biosphere" [2]. Most of the flora of the rare biosphere includes fungi, hence known as "mycobiome." There is also a component of the microbiome (virome) and parasites. A vast majority of bacteria in the oral cavity has been categorized in the Human Microbiome Project (HMP). HMP has also developed an organ-specific database (HOMD) [http://www.homd.org] [3]. It is a comprehensive information database on the approximately 700 prokaryote species present in the human oral cavity. HOMD is based on a curated 16S rRNA gene-based provisional naming scheme. Approximately 700 species are listed in the HOMD, out of which only 51 percent are officially named, 13% are not named (but cultivated), and 28% are known only as uncultivated phylotypes [4].

The main aim of the database was to understand the diversity of the microbes that colonize the different niches in human body and to evaluate the interplay between health and disease with the microbial composition. This can lead to development of tools for microbial research and creating public repositories [5]. Since then, it has culminated into a global collaborative effort: "International Human Microbiome Consortium." The first studies gathered data on the microbial diversity across various anatomical niches in normal and numerous pathological states. These are now opening avenues for human microbiota-driven research into various infectious and even noncommunicable diseases. The hypotheses-driven research has shown many disease states to be associated with local and distant microbial community structure and function. The microbial composition has been associated with alteration of mucosal barrier function, energy metabolism, utilization, and interaction with pathogenic microorganisms impacting the host [6].

6.2 The Microbiome of Oral Cavity

The two critical features in normal host-microbiome are:

A. Functional Core: The healthy human microbiome shows the presence of a "functional core" which constitutes a distinct core group of microbial gene families, transcriptional pathways, and metabolome consistent with health [7]. These functional pathways involve basic housekeeping genes for survival of commensal microbes, genes involved in host-microbial cross talk such as microbial adhesion to host cells, production of nutrient compounds, metabolism, and immune modulation. The functional core of the microbiome is also body site-specific, e.g., vitamin K and short-chain fatty acid production by the gut microbiota [8].

B. Alpha and Beta Diversity: Microbiome diversity is an essential component of functional core. Alpha diversity refers to the different microorganisms present at a single site/niche/sample. Beta diversity is the measure of similarity and variation present in microbial communities of two different samples/sites.
C. Dysbiosis: The alterations in microbial composition which results in the disturbances in the microbial community culminating in the ability to function and sustain optimally.

6.3 Evolutionary Concepts

6.3.1 Predentate Microbiome

Colonization of oral mucosa at birth is influenced by many factors, such as a combination of different mother microbiome sources, including the gut, vaginal, skin microbiome, the mode of delivery of the infant, and breast milk. The oral microbial colonization of C-section babies resembles that of the mother's skin, e.g., Propionibacterium, Corynebacterium, and Streptococcus. The oral microbiome of vaginal delivery resembles the colonization from the mother's vagina, characteristically Prevotella, Sneathia, and Lactobacillus genera [9]. Slackia exigua was detected only in infants delivered by C-section. Infants delivered vaginally were harboring more species of bacterial taxa [10, 11].

The newborn oral cavity seems to be rapidly dominated by Streptococcus species (*S. mutans, S. epidermidis, S. salivarius*) and Fusobacterium. There is a predominantly high abundance of Streptococcus species, as reported in many studies. The reasons could be twofold: (1) Streptococcus species has the ability to adhere to epithelial cells, and (2) Streptococcus spp. is one of the dominant bacterial groups found in human breastmilk [12]. Among Streptococcus species, Streptococcus salivarius is the most often found species in the oral cavity of the newborn, and its abundance reaches the highest at 3 months of age [10, 13] accounting for 10–15% of the total Streptococcal species. The metabolic activity on dietary oligosaccharides in breast milk or formula milk by Streptococcus makes the environment conducive to other commensals' growth [14].

In addition to the early colonizers, bacterial genera *Gemella*, *Granulicatella*, *Haemophilus*, and *Rothia* present at 3–6 months of age with more than 1% abundance; and their abundance further increases with time [15]. In the first 3–4 months of an infant, *E. coli*, *Staphylococcus*, and *Pseudomonas* are predominantly present. Lactic acid-producing bacteria such as *L. crispatus*, *L. gasseri*, and *Veillonella* start colonization after tooth eruption.

6.3.2 Microbiome in Deciduous Dentition

With the eruption of teeth, the oral microbiome undergoes evolution into a different co-community of microorganisms. This is also primarily due to the presentation of a new adhesion surface—teeth. According to Caufield et al., this was termed "Window of Infectivity" as this stage was thought to be the initiation stage for colonization of cariogenic streptococcus species like *S. mutans*. Nevertheless, evidence now shows the presence of *S. mutans* in edentulous children also [16].

The primary dentition is a stage of the most significant expansion in phylogenetic and functional diversity in the oral microbiome. This is due to change in adhesion surfaces, cellular manipulation, and biofilm formation, and change or alteration in microbial habitats that drive different species to colonize the niche [17].

An established community in the deciduous dentition is:

Actinobacteria—Actinomyces, Rothia, Corynebacterium

Bacteroidetes—Prevotella, Porphyromonas, Capnocytophaga, Cloacibacterium

Firmicutes—Granulicatella, Streptococcus, Lactobacillus, Veillonella, Selenomonas, Peptostreptococcus, Oribacterium, Mogibacterium

Fusobacteria—Fusobacterium, Leptotrichia, Chitinbacter, Neisseria, Shewanella

Proteobacteria—Actinobacillus, Haemophilus, Aggregatibacter, Acinetobacter

The diet and environment determine the mature oral microbiome. This interaction increases significantly in diversity which is also responsible for colonization of potentially cariogenic taxa. This stage is also critical for the emergence of oral diseases, with sugar consumption and fermentation by potentially cariogenic taxa (along with acidic pH) as the principal contributing factors [14].

6.3.3 Microbiome in Mixed Dentition

In mixed dentition, deciduous teeth exfoliate, and permanent teeth erupt, thereby modifying the ecosystem. Changes in the surface enamel for biofilm formation [18] and residence time in the oral cavity alter the deciduous and permanent teeth niche, affecting the microbial composition [19].

The core microbiome consisted of 19 genera as detected in many studies [19–23]:

Proteobacteria—Aggregatibacter, Cardio bacterium, Lautropia, Campylobacter, Haemophilus, Neisseria

Firmicutes—Gambella, Granulicatella, Abiotrophia, Streptococcus

Actinobacteria—Actinomyces, Corynebacterium, Rothia, Gemella

Fusobacteria—Fusobacterium, Leptotrichia

Bacteroidetes—Porphyromonas, Prevotella, Capnocytophaga

Among the changes in saliva and supragingival plaque in mixed dentition, Firmicutes were more abundant in saliva, and Fusobacteria and Actinobacteria were more abundant in plaque [24].

6.3.4 Microbiome in Permanent Dentition

This phase comprises of microbiome seen in adolescence to adulthood with all erupted permanent teeth. In adult life, the resident microbes are reasonably stable due to the complex dynamic interactions between inter-bacterial and host-bacterial interactions. This is termed "microbial homeostasis" [25]. This implies that the oral cavity is naturally and gradually colonized by different microbial taxa with progressing age, amidst maintaining a healthy status.

Mager et al. in 2003 proved that distinct ecological niches like tooth colonizing species and soft tissue colonizing species exist. Most of the salivary microflora was similar to that found on the dorsum and lateral surface of the tongue. At the genus level, analysis of saliva and supragingival plaque represented 318 different genera. Saliva harbored 185 different types of microflorae as compared to 267 different microbiomes detected in plaque samples. Fifty percent of the salivary microflora (phylum) consisted of 3 genera—*Prevotella, Streptococcus, Veillonella.* Fifty percent of plaque samples comprised of *Streptococcus, Veillonella, Corynebacterium, Actinomyces, Fusobacterium,* and *Rothia* [20, 21].

6.3.5 Niche and Inter-Subject Variation Within the Oral Microbiome

The oral cavity constitutes a unique ecological system with many sub-niches of diverse environmental characteristics reflected in the dynamics of the populating microbiota [23]. These sub-niches are primarily buccal mucosa, palatal mucosa, tongue dorsum, shedding soft surfaces, and non-shedding hard surfaces, i.e., essentially dental surfaces bearing supragingival or subgingival plaque and saliva. The dental plaque shows the highest microflora in species richness and diversity, and the mucosal microbiome is the lowest [21, 23, 26]. The salivary microbiome is more similar to mucosal surfaces, primarily derived from the microbiota shredded from the mucosa, but it also contains microbiota from dental plaque. The significant differences between plaque and mucosal/salivary microbiomes are attributable to widely variable microenvironments. Non-shedding tooth surfaces allow more significant time for microbial community maturation and therefore espouse more complex microbiomes. Other factors accounting for such differences include differences in adhesion molecules, inter-species interactions, host genetics, variation in random events of microbial dispersal or community disruption [27].

Inter-subject variation is a significant driver of microbiome diversity in the oral niches as elsewhere. Ethnic or geographic signatures on oral microbiota are documented, significantly impacting saliva on subgingival niches than the supragingival niche.

In a seminal study, Bashan et al. [28] asked if microbial community dynamics could depend on host attributes (genetic makeup and lifestyle) or host-independent microbial factors such as metabolic pathways and inter-species interactions. They demonstrated a "universal" microbiome dynamic in gut and saliva in health,

Table 6.1 The normal oral microbiome seen in the various ecological niche in oral cavity

Different sites in the oral cavity	Bacterial species commonly inhabiting
Tongue	*Porphyromonas gingivalis, Veillonella atypica, Aggregatibacter actinomycetemcomitans, Prevotella intermedia, Capnocytophaga Enterococcus faecalis, Eikenella corrodens*
Gingival crevice	*Fusobacterium, Prevotella, Porphyromonas, Streptococcus mitis, Streptococcus sanguinis, Propionibacterium acnes, Leptotrichia buccalis, Actinomyces odontolyticus, Veillonella parvula*
Dental plaque and tooth	*Actinomyces, Rothia, Microbacterium, Mycobacterium, Propionibacterium, Corynebacterium, Bifidobacterium, S mutans, Actinomyces, Eubacterium, Peptostreptococcus*
Posterior part of oral cavity/oropharynx	*Streptococcus viridans, Haemophilus influenzae, Neisseria species, Staphylococcus, Streptococcus pyogenes, Streptococcus salivarius, Streptococcus mutans, Haemophilus parainfluenzae*

indicating that inter-subject variability in the gut and oral microbiome structure stems mainly from differences in microbial assembly. The most common microorganisms present in most of the individuals at various sites in oral cavity are listed in Table 6.1.

6.4 Mycobiome of the Oral Cavity

The principal fungi present in the oral cavity is the Candida species. The four most common oral mycobiome consist of *Candida*, *Aspergillus*, *Fusarium*, and *Cryptococcus* [29]. The most abundant *Candida* species reported were *Albicans* (in 40% of the subjects), followed by *C. parapsilosis* (15%), *C. tropicalis* (15%), *C. khmerensis* (5%), and *C. metapsilosis* (5%). The distribution and profile of fungal species in the oral cavity of healthy individuals are complex and similar to that of the oral bacterial microbiome concerning the number of species identified.

Aspergillus, *Fusarium*, and *Cryptococcus* isolates, which are pathogenic, are reported in the oral cavity of healthy individuals in a few studies. It is possible that in healthy individuals the pathogenicity of these fungi is downregulated by other fungi in the oral mycobiome and a functional immune system. Interdependent relationships may exist between components of the oral mycobiome, which could be put to use during therapeutic modalities. Studies have also shown that around 60 fungal genera present in the environment like plants, soil, and air were also seen in the oral cavity. The logical explanation is that they are implanted from food and due to mouth breathing. Many nonculturable fungal genera that were not previously detected are now being identified, which may be responsible for antimicrobial resistance or can explain the proliferation of other pathogenic species.

6.5 Methods for Determining Oral Microbiome

6.5.1 Culture Technique

Initially, the microbial taxa were identified using various technologies such as microscopy, biochemical and phenotypic tests, fermentation tests, and antibiotic sensitivity that has culture as the main mode. However, these methods are time-consuming, expensive, and labor-intensive, and not as comprehensive as the other modern techniques [30].

6.5.2 Gel-Based Technologies

These are community profiling methods based on PCR, where specific primers are used for amplification and analysis techniques, such as denaturing gradient gel electrophoresis (DGGT) and temperature gradient gel electrophoresis (TGGE), which are used for separating and identifying DNA [31]. Restriction fragment length polymorphism (RFLP) is used as a tool for genome mapping [32].

6.5.3 Microbial Analysis Using Sequencing Methods

The current era has seen a revolution of sorts in molecular biology, computational biology, biotechnology, and bioinformatics. This has resulted in a new subdiscipline of "omics": open-ended detection of biological entities viz. genes (genomics), proteins (proteomics), messenger RNA (transcriptomics), and metabolites (metabolomics) which can have great impact on practice and medicine [33].

The advancement towards time-effective and large-scale molecular "sequencing" of nucleic acids has grown potentially in the last decade, with the progression to highly parallelized next-generation sequencing (NGS) [34]. Since the first report of sequencing of bacterial genomes in 1995 [35, 36], sequences from 50 different bacterial phyla have been studied [37]. NGS technologies have many advantages over the previous technologies due to the speed, vast data handling and sharing.

Sequencing methods may be listed as follows.

6.5.4 First Generation or Sanger Sequencing

"Sanger" sequencing is comprised of the "first generation" of nucleic acid sequencing since its development in 1975–1977 [38]. It is based on chain termination and strand elongation by DNA polymerase. The first automated sequencer was launched in 1986, which used slab gel electrophoresis. Initially the number of sequences per sample was limited as it was cost and labor intensive. This was further improvised on with automated reloading of the slab gels. Despite these, Sanger sequencing with sequence lengths of 1000 base pairs was the dominant sequencing method till 2005, heralding the next-generation sequencing era.

6.5.5 Next-Generation or High-Throughput Sequencing (NGS)

The 2000s saw the advent of several high-throughput sequencing platforms. Roche 454, introduced in 2004, was the first commercial next-generation sequencing which was followed by Illumina in 2006. These used parallel DNA sequence analysis which yields millions of sequences reads per sample, overcoming the limitations of Sanger sequencing. However, these platforms had various specific shortcomings [39].

6.5.6 The "Second-Generation" Sequencing Platforms

Among the many second-generation sequencing platforms, the prominent ones are Roche 454 which is an emulsion PCR which uses pyrosequencing (shot gun sequencing procedure). It reads an average of 400–500 base pairs. The ABI/SOLiD also uses emulsion PCR (sequencing by ligation) with an average read of 75 base pairs. Illumina uses solid phase amplification with one DNA fragment per cluster. It has an average read of 300 base pairs (sequencing by synthesis). Each of these platforms has different pros and cons, and the scenario was dominated by the Illumina platform [40].

6.5.7 Third-Generation Sequencing

The "third-generation" sequencers have been developed that can be used to study genomes and transcriptomes at a high resolution. These overcome the distinct disadvantages of the NGS like relatively short reads which can result in gaps and mis-assembly. The PacBio platform, which sequences single molecules based on phosphor-linked fluorescent nucleotides (sequencing-biosynthesis) [41], is currently the most popular single-molecule real-time (SMRT) sequencing modality. Apart from long reads, it also enables the detection of DNA methylation. Nanopore sequencing platforms are also being developed which have multiple recognition points that determine DNA sequences and can be customized in small laboratories. The common nanopore sequencers are provided by Oxford Nanopore, GridION, and the portable MinION [42]. These are based on the passage of DNA molecules through small "nano" scale pores with measurement of change in electrical conductivity enabling detection of individual nucleotide bases [43].

References

1. Woese CR, Kandler O, Wheelis ML (1990) Towards a natural system of organisms: propossal for the domains Archaea, Bacteria, and Eucarya. Proc Natl Acad Sci U S A 87(12):4576–4579
2. Huse SM, Dethlefsen L, Huber JA et al (2008) Exploring microbial diversity and taxonomy using SSU rRNA hypervariable tag sequencing. PLoS Genet 4(11):e-1000255
3. Dekaboruah E, Suryavanshi MV, Chettri D et al (2020) Human microbiome: an academic update on human body site specific surveillance and its possible role. Arch Microbiol 202(8): 2147–2167
4. HOMD: Human Oral Microbiome Database (nd) http://www.homd.org/?name+HOMD
5. Turnbaugh PJ, Ley RE, Hamady M et al (2007) The human microbiome project: exploring the microbial part of ourselves in a changing world. Nature 449(7164):804–810
6. Cani PD, Plovier H, Van Hul M et al (2016) Endocannabinoids-at the crossroads between the gut microbiota and host metabolism. Nat Rev Endocrinol 12(3):133–143
7. Martiny JBH, Jones SE, Lennon JT et al (2015) Microbiomes in light of traits: a phylogenetic perspective. Science 350:9323–9328
8. Flint HJ, Scott KP, Duncan SH, Louis P, Forano E (2012) Microbial degradation of complex carbohydrates in the gut. Gut Microbes 3(4):289–306
9. Dominguez-Bello MG, Costello EK, Contreras ME et al (2010) Delivery mode shapes the acquisition and structure of the initial microbiota across multiple body habitats in newborns. Proce Natl Acad Sci USA 107(26):11971–11975
10. Hegde S, Munshi AK (1998) Influence of the maternal vaginal microbiota on the oral microbiota of the newborn. J Clin Pediatr Dent 22(4):317–321
11. Lif Holgerson P, Harnevik L, Hernell O et al (2011) Mode of birth delivery affects oral microbiota in infants. J Dent Res 90(10):1183–1188
12. Xiao J, Fiscella KA, Gill SR (2020) Oral microbiome: possible harbinger for children's health. Int J Oral Sci 12(1):1–13
13. Boix-Amorós A, Collado MC, Mira A (2016) Relationship between milk microbiota, bacterial load, macronutrients, and human cells during lactation. Front Microbiol 7:1–9
14. Gomez-Arango LF, Barrett HL, McIntyre HD et al (2017) Contributions of the maternal oral and gut microbiome to placental microbial colonization in overweight and obese pregnant women. Sci Rep 7(1):1–10
15. Dzidic M, Collado MC, Abrahamsson T et al (2018) Oral microbiome development during childhood: an ecological succession influenced by postnatal factors and associated with tooth decay. ISME J 12(9):2292–2306
16. Gizani S, Papaioannou W, Haffajee AD et al (2009) Distribution of selected cariogenic bacteria in five different intra-oral habitats in young children. Int J Paed Dent 19(3):193–200
17. Mason MR, Chambers S, Dabdoub SM et al (2018) Characterizing oral microbial communities across dentition states and colonization niches. Microbiome 6(1):67–77
18. De Menezes Oliveira H, Oliveira MA, Torres CP, Gomes-Silva JM et al (2010) Microstructure and mineral composition of dental enamel of permanent and deciduous teeth. Microsc Res Tech 73(5):572–577
19. Shi W, Qin M, Chen F (2016) Supragingival microbial profiles of permanent and deciduous teeth in children with mixed dentition. PLoS One 11(1):1–15
20. Aas JA, Paster BJ, Stokes LN et al (2005) Defining the normal bacterial flora of the oral cavity. J Clin Microbiol 43(11):5721–5732
21. Keijser BJF, Zaura E, Huse SM et al (2008) Pyrosequencing analysis of the oral microflora of healthy adults. J Dent Res 87(11):1016–1020
22. Wang WL, Xu SY, Ren ZG et al (2015) Application of metagenomics in the human gut microbiome. World J Gastroenterol 21(3):803–814
23. Xu X, He J, Xue J et al (2015) Oral cavity contains distinct niches with dynamic microbial communities. Environ Microbiol 17(3):699–710

24. Shi W, Tian J, Xu H et al (2018) Distinctions and associations between the microbiota of saliva and supragingival plaque of permanent and deciduous teeth. PLoS One 13(7):1–17
25. Marsh PD, Percival RS (2006) The oral microflora - friend or foe? Can we decide? Int Dent J 56(4):233–239
26. Huttenhower C, Gevers D, Knight R et al (2012) Structure, function and diversity of the healthy human microbiome. Nature 486(7402):207–214
27. Stegen JC, Lin X, Konopka AE et al (2012) Stochastic and deterministic assembly processes in subsurface microbial communities. ISME J 6(9):1653–1664
28. Bashan A, Gibson TE, Friedman J (2016) Universality of human microbial dynamics. NHS Public Access 534(7606):259–262
29. Ghannoum MA, Jurevic RJ, Mukherjee PK et al (2010) Characterization of the oral fungal microbiome (mycobiome) in healthy individuals. PLoS Pathog 6(1):e1000713
30. Parahitiyawa NB, Scully C, Leung WK et al (2010) Exploring the oral bacterial flora: current status and future directions. Oral Dis 16(2):136–145
31. Nishigaki K, Naimuddin M, Hamano K (2000) Genome profiling: a realistic solution for genotype-based identification of species. J Biochem 128(1):107–112
32. Deng W, Xi D, Mao H et al (2008) The use of molecular techniques based on ribosomal RNA and DNA for rumen microbial ecosystem studies: a review. Mol Biolol Rep 35(2):265–274
33. Alcaraz LD, Belda-Ferre P, Cabrera-Rubio R et al (2012) Identifying a healthy oral microbiome through metagenomics. Clin Microbiol Infect 18(4):54–57
34. Margulies M, Egholm M, Altman WE et al (2005) Genome sequencing in microfabricated high-density picolitre reactors. Nature 437(7057):376–380
35. Fleischmann RD, Adams MD, White O et al (1995) Whole-genome random sequencing and assembly of Haemophilus influenzae Rd. Science 269(5223):496–512
36. Fraser CM, Gocayne JD, White O et al (1995) The minimal gene complement of Mycoplasma genitalium. Science 270(5235):397–403
37. Land M, Hauser L, Jun SR et al (2015) Insights from 20 years of bacterial genome sequencing. Funct Integrat Genom 15(2):141–161
38. Sanger F, Nicklen S, Coulson AR (1977) DNA sequencing with chain-terminating inhibitors. Proc Natl Acad Sci U S A 74(12):5463–5467
39. Schirmer M, Ijaz UZ, D'Amore R (2015) Insight into biases and sequencing errors for amplicon sequencing with the Illumina MiSeq platform. Nucleic Acids Res 43(6):e37
40. Greenleaf WJ, Sidow A (2014) The future of sequencing: convergence of intelligent design and market Darwinism. Genome Biol 15(3):4–6
41. Hodkinson BP, Grice EA (2015) Next-generation sequencing: a review of technologies and tools for wound microbiome. Adv Wound Care 4(1):50–58
42. Eisenstein M (2012) Oxford nanopore announcement sets sequencing sector abuzz. Nat Biotechnol 30(4):295–296
43. Lu H, Giordano F, Ning Z (2016) Oxford nanopore MinION sequencing and genome assembly. Genom Proteom Bioinform 14(5):265–279

Oral Microbiome in Oral Squamous Cell Carcinoma

7

Supriya Kheur, Aneesha Acharya, Mohit Kheur, Vinay Jain, and Vinay K. Hazarey

Abstract

In recent years, studies have shown that like many other diseases, oral cancer initiation and progression is influenced by a cocktail of host immunity, microenvironment, and resident microbiota. Though bacteria like *H. pylori*, Fusobacterium, and Porphyromonas and viruses like EBV and HPV are established risk factors for cancers in various sites of human body, studies linking microbiome and oral cancers are still in their preliminary stages. This chapter provides an insight on the links between dysbiotic microbiota and oral cancer,

S. Kheur (✉)
Department of Oral Pathology and Microbiology, D.Y. Patil Dental College and Hospital, Dr. D.Y. Patil Vidyapeeth, Pune, Maharashtra, India
e-mail: supriya.kheur@dpu.edu.in

A. Acharya
Department of Periodontics, Dr. D.Y. Patil Dental College and Hospital, Dr. D.Y. Patil Vidyapeeth, Pune, Maharashtra, India
e-mail: aneesha.acharya@dpu.edu.in

M. Kheur
Department of Prosthodontics, Rangoonwala College of Dental Sciences and Research Centre, Pune, Maharashtra, India
e-mail: mkheur@gmail.com

V. Jain
Department of Prosthodontics, University of Tennessee Health Sciences Centre College of Dentistry, Memphis, TN, USA
e-mail: vjain@uthsc.edu

V. K. Hazarey
Adjunct Faculty, Dr. D.Y. Patil Dental College and Hospital, Dr. D.Y.Patil Vidyapeeth, Pune, Maharashtra, India
e-mail: vinay.hazarey@gmail.com

S. Routray (ed.), *Microbes and Oral Squamous Cell Carcinoma*,
https://doi.org/10.1007/978-981-19-0592-6_7

with a particular focus on their pathogenesis, immune responses, metabolism, and strategies to target the microbiome for cancer prevention.

Keywords

Oral cancer · Microbiome · Pathogenesis · Immune response · Cancer prevention

7.1 Introduction

The early detection of OSCC at initial stages is crucial and requires less radical therapy, thereby improving quality of life. The primary risk factors for oral cancers are tobacco consumption, *Areca* nut, alcohol, chronic irritation, and poor oral hygiene. Other possible risk factors include viral infections, infections with candida species, periodontitis, chronic bacterial infection, and inflammation [1, 2].

Oral carcinogenesis can be broadly categorized into three phases of development, namely (1) Initiation, (2) Promotion, and (3) Progression. Initiation is brought about by potential carcinogenic agents that introduce oncoproteins in the target cell, which further causes activation of protooncogenes and suppression of tumor suppressor genes by altering the genetic code of the target cells. Promotion involves the growth of the mutated cell through upregulation of angiogenic factors, growth factors, and cytokines. The tumor microenvironment mimics chronic inflammatory lesions causing extracellular matrix destruction and altered immune response. In the final phase of carcinogenesis, the malignant cells invade the local and distant tissue. The process of molecular alterations causing carcinogenesis continues in the new site.

7.2 Dysbiosis of Oral Microbiome and Carcinogenesis

An emerging concept in cancer biology implicates the microbiome as an influential factor that can modulate carcinogenesis. The oral cavity is a habitat to a diverse microbial community in a disease-free state. Dysbiosis, characterized by a general loss of microbial diversity, is marked by increased proliferation of pathobiont microorganisms. Notably, several viruses like EBV, HBV, and HPV are associated with cancers. Among the various bacteria with tumor-promoting/initiating properties, *H. pylori* were among the first to be associated and are now recognized as a class-1 carcinogen [3]. The microbial taxa present at the cancer site may also reflect local infection existing in neoplastic tissue. William Coley, also famously known as "the father of immunotherapy," pioneered concomitant bacterial infection of tumors. He identified the presence of *Streptococcus pyogenes* in soft tissue sarcoma. The current consensus is that the oral microbiome may play an essential role in carcinogenesis by either producing and augmenting certain carcinogens or mimicking an inflammatory response.

Further, known carcinogens like areca and betel nut, chewing tobacco, and alcohol can alter the normal microflora causing dysbiosis. Tobacco consumption in both smokeless and smoking forms and alcohol consumption and dietary factors can also contribute to dysbiosis leading to malignant transformation [4].

Multiple factors contributing to bacterial growth in tumors [5] include:

1. Hypoxic environment in solid tumors.
2. Presence of bacterial nutrients such as purines.
3. Release of factors like aspartate, serine, citrate, galactose by cancer cells.
4. Neo-angiogenesis and formation of "leaky" new blood vessels.

Recently, Al-Hebshi et al. developed a "passenger turning driver" concept/theory explaining the possible role of the oral microbiome in oral carcinogenesis [6]. They proposed that unlike cancers seen in other parts of the body, oral microbiome is not the driver of oral carcinogenesis. This theory proposes that the incipient tumor represents a passenger event due to the growth of certain bacterial species that thrive in the tumor microenvironment, e.g., Fusobacteria, Gram-negative anaerobic rods, anaerobic cocci, and *C. Albicans*. With the proliferation of the tumor cells, this intratumor-microbiome expresses various pro-inflammatory cytokines and virulence factors, turning it into an altered "Driver" microbiome—this further causes progression into oral cancer.

7.2.1 Bacterial Infections Associated with Oral Carcinogenesis

Bacterial species increased in OSCC is *P. melanogenic* [7], *Streptococcus anginosus*, *Veillonella parvula*, *Prevotella pallens*, *Porphyromonas endodontalis*, *Prevotella nanceiensis*, *Dialisten* sp., *Campylobacter ureolyticus*, *Fusobacterium* sp., *P. nigrescens*, *Neisseria bacilliformis*, *Peptostreptococcus anaerobius*, and *Neisseria subflava.* The alterations in oral bacteria as observed in multiple studies have been summarized in Table 7.1 [8, 9]. Five bacteria (*Fusobacterium periodontium*, *Streptococcus constellatus*, *Hemophilia influenza*, *Prevotella micro* and *Fusobacterium alocis)* were significantly over-represented in OSCC patients, and their relative abundance microbiota has been proposed for monitoring OSCC [10].

Chronic inflammation is the most crucial factor induced by the oral microbiome to promote oral carcinogenesis. Periodontal pathologies also produce a constant state of inflammation. The periodontal-pathogenic bacterial colonization is highly correlated with inflammation and subsequent oncogenesis. Many parameters in carcinogenesis like cell proliferation, angiogenesis, and oncogenesis may be caused and facilitated by inflammatory mediators. Thus, chronic inflammation is an essential hallmark of the development of any cancer. IL-1b, IL-6, and TNF-a, among others, can trigger the Toll-like receptors (TLR)-MyD88 signaling pathway in leukocytes and epithelial cells. These pathways, in turn, mediate the activation of NF-KB, STAT-3, and AP-1 transcription factors. These factors, in turn, upregulate the antiapoptotic genes like c-Myc, MC-1, cyclin-D, BCL-2, C-Flip, and Survivin [11].

Table 7.1 Alterations in bacteria identified/studied in oral cancer (Chattopadhyay I)

Name of the phylum	Name of the bacteria	Type of sample studied
Reduced		
Actinobacteria		
	Actinomyces odontolyticus	Tissue biopsy
	Rothia mucilaginosa	Tissue biopsy
Firmicutes		
	Streptococcus mitis	Tissue biopsy
	Veillonella parvula	Oral rinse
Bacteroidetes		
	Capnocytophaga	Oral mouthwash
Fusobacteria		
	Leptotrichia	Saliva
Increased		
Fusobacteria		
	F. naviforme	Oral swab
	F. nucleatum	Saliva
	F. periodonticum	Tissue biopsy
Firmicutes		
	Streptococcus oralis	Tissue biopsy and saliva
	Streptococcus salivarius	Tissue biopsy and saliva
	Streptococcus parasanguinis	Oral swab and saliva
	Staphylococcus aureus	Oral swab and saliva
	Streptococcus anginosus	Oral swab and saliva
	Dialister pneumosintes	Oral rinse
	Lactobacillus gasseri	Tissue biopsy and saliva
	Lactobacillus fermentum	Tissue biopsy and saliva
	Lactobacillus rhamnosus	Tissue biopsy and saliva
Bacteroidetes		
	Porphyromonas gingivalis	Oral wash
	Prevotella melaninogenica	Tissue biopsy and saliva
	Capnocytophaga gingivalis	Oral swab
	Treponema forsythia	Oral rinse
	Prevotella salivae	Tissue
Proteobacteria		
	Aggregatibacter	Oral wash
	Pseudomonas aeruginosa	Tissue biopsy

The inflammatory microenvironment is also caused and sustained by secreting reactive oxygen species (ROS) through NAPDH oxidase. It catalyzes the superoxide anion (O_2^-), which causes the production of hydrogen peroxide via superoxide dismutase. Reactive Nitrogen Species also generates Nitric Oxide (NO), which can be converted into nitrogen dioxide (NO_2), peroxynitrite (ONOO−), and dinitrogen trioxide (N_2O_3). Increased ROS and RNS with subsequent increased expression of NADPH oxidase are seen consistently in oral cancer. Most common bacteria

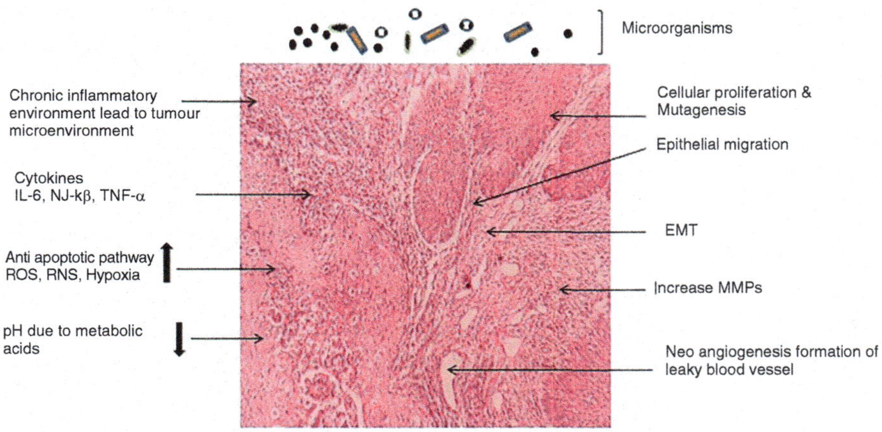

Fig. 7.1 Role of microbiome in oral carcinogenesis

associated with ROS and NOS include *Streptococcus oralis*, *Streptococcus mitis*, *Streptococcus sanguinis*, *Streptococcus gordonii*, *Streptococcus oligofermentans*, *Lactobacillus fermentum*, *Lactobacillus jensenii*, *Lactobacillus acidophilus*, and *Bifidobacterium adolescentis*. Similarly, *Porphyromonas gingivalis*, *Fusobacterium nucleatum*, *Prevotella intermedia*, and *Aggregatibacter actinomycetemcomitans* can produce hydrogen sulfide (H_2S), methyl mercaptan (CH_3SH), dimethyl sulfide $(CH_3)_2S$, and dimethyl disulfide (CH_3SSCH_3). These compounds can cause mutations and genomic stability [12].

Bacterial species like Streptococcus, Lactobacillus, and Bifidobacterium produce acids like lactic acid, acetic acid, butyric acid, isobutyric acid, isovaleric acid, and isocaloric acids. These acids lower the pH and contribute towards the formation of an acidic and hypoxic environment. The tumor microenvironment also plays an active role in the increased cohabitation of microorganisms. Aggressive malignancies are often associated with the heightened concentration of lactic acid, LDH protein expression, hypoxia, and acidity [13]. Other contributing factors include bacterial species like Streptococci and Neisseria and fungal species like Candida, which possess the enzyme alcohol dehydrogenase (ADH). This converts alcohol to carcinogenic acetaldehyde, which causes mutations by forming adducts with DNA.

Thus, overall microbial products create a chronic inflammatory microenvironment promoting carcinogenesis by:

1. Activation of transcription factor NF-KB.
2. Bacterial products like lipopolysaccharides (LPS), various enzymes like fibrinolysin, collagenases, proteases, phospholipase, etc., and the by-products of metabolism like reactive oxygen species nitrogen species, various fatty acids, and some metabolites like malondialdehyde-4-hydroxy −2-nominal. These induce inflammatory responses.
3. These, in turn, cause DNA mutations and alterations (Fig. 7.1).

The two most extensively studied bacteria for their role in oral carcinogenesis are *Porphyromonas gingivalis* (*P. gingivalis*) and *Fusobacterium nucleatum* (*F. nucleatum*)

7.2.1.1 *P. gingivalis* and Its Role in Oral Cancer

P. gingivalis plays an important role in oral carcinogenesis. It causes an increase in cell proliferation by modulating the apoptotic pathway. After infection with *P. gingivalis*, gingipains cause STAT-3 activated IL-6, promoting cyclin D1 expression. The JAK-1/Akt/STAT-3 signaling causes downregulation of Bad with alteration of BcL2/Bax ratio. It also downregulates ATP ligation of P2X7 receptors, thereby preventing apoptosis [14]. It also destroys collagen IV and V of the extracellular matrix and increases the expression of MMP-9, which ZEB1 or TLR-2 may activate. An increase in MMP-9 expression has been seen with the alteration of p38 MAPK and ERK1/2, and NF-KB signal pathways are activated by PAR2 and PAR4. Its MMP-1,2,9,10 increased, and IL-8 could also be increased, causing extracellular matrix degradation. Upregulation of ZEB1, or SNAIL/Slug/TWIST transcription factors, reduces epithelial expression in cancer cells, thus laying an important role in EMT of oral epithelial cells. The lipopolysaccharide (LPS) component of *P. gingivalis* causes the release of inflammatory pro-cytokines by Toll-like receptor-4 (TLR-4) and the NF-KB pathway [15].

7.2.1.2 *F. nucleatum* and Its Role in Oral Cancer

F. nucleatum infection in oral epithelial cells promotes increased cellular proliferation and migration via stimulation of Etk/BMX, S6 kinase p70, and Rho A kinase. Fusobacterium also activates mitogen-activated protein kinase p38. This results in increased expression of MMP-9 and MMP-13 with upregulation of IL-1b active form. It also plays an essential role in EMT by upregulating mesenchymal markers, including E cadherin and Vimentin and partial EMT genes [16].

7.2.2 Fungal Infections Associated with Oral Carcinogenesis

Several studies exhibit an association of fungal infections with oral cancer and malignant transformation of Oral Potentially Malignant disorders into Oral cancer. Oral candidal carriage is higher and more diversified in oral cancer patients as compared to regular counterparts. The most common yeasts other than Candida species are Rhodotorula, Geotrichum, Pneumocystis, and Fonsecaea [17]. These fungi can alter the cancer microenvironment. They also cause an increase in other pathogenic oral microorganisms leading to increased aggressiveness. *C. albinos* have been associated with a form of oral potentially malignant lesion known as Candidal leukoplakia. During infections with *C. albicans*, the organism shifts from budding yeast to hyphal form. Recently, a new *C. albicans* toxin known as candidalysin has been found encoded by the ECE1 gene. This gene is associated with conversion into hyphal form and host cell adhesion. Candidalysin also damages the epithelial cells of the mucosa. This triggers the release of pro-inflammatory

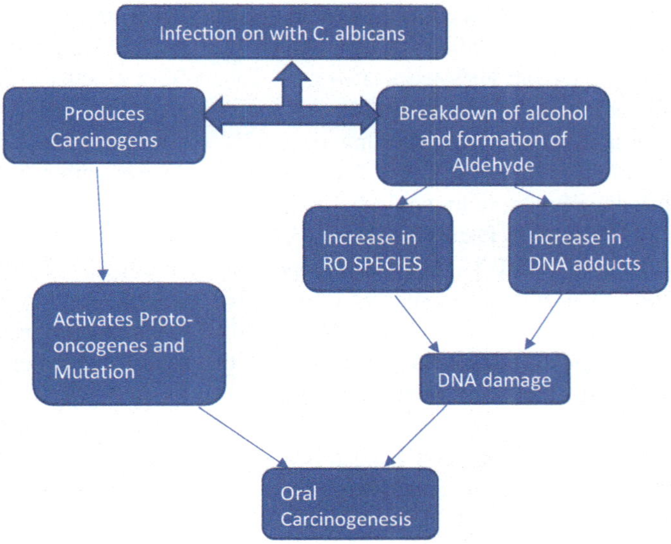

Fig. 7.2 Role of *Candida albicans* in oral carcinogenesis

interleukins like IL-1α, IL-1β, IL-6, IL-23, granulocyte-macrophage colony-stimulating factor (GM-CSF), CCL20, and β-defensins to be released [18]. Some authors also report that Candida association with Oral Cancer is also due to its ability to metabolize acetaldehyde from ethanol by the enzyme Alcohol dehydrogenase (ADH), which can cause carcinogenesis [19].

The nitrosamine compounds produced by Candida are shown to produce NBMA (N-nitroso-benzylmethylamine) from salivary precursor molecules [20, 21]. The role of *C. albicans* in the promotion of oral carcinogenesis is depicted in Fig. 7.2.

7.2.3 Viral Infections Associated with Oral Carcinogenesis

Human Papillomavirus (HPV): 25–30% of oral cancer can be attributed to HPV, and HPV type 16 is the most critical subtype in oral carcinogenesis. The involvement of HPV in oral carcinogenesis was first proposed when HPV-associated cytopathic effects—Koilocytes—were noted in the epithelial cells of the oral cavity [22]. Most studies focus on HPV-16 and 18 strains which are considered oncogenic. The HPV genome has specifically E6 and E7 oncoproteins under the control of E2 and E1 inhibitory genes. These genes get blocked upon integration, thereby causing unchecked transcription of E6 and E7. The E6 can inactivate p53 through the ubiquitin-dependent degradation pathway. p53 is an essential tumor suppressor gene associated with oral carcinogenesis. E7 can phosphorylate Rb protein (another tumor suppressor protein), causing its degradation. This subsequently leads to E2F activation, which causes the cellular proliferation [23].

The association of HPV is established with oropharyngeal cancers, and its role in oral cancer is more in young male adults and includes non-smokers and non-drinkers. In general, HPV-associated oral cancers have a better prognosis and are generally amenable to treatment [24].

EBV is a ubiquitous virus infecting over 95% of people. It is also the first recognized human oncogenic virus. The virus is associated with any neoplasms like Burkitt's lymphoma, immunoblastic lymphomas, nasopharyngeal cancers, and OSCC [25]. The various factors implicated in EBV-related oncogenesis are latent membrane protein (LMP-1), EBV-determined nuclear antigen (EBNA)-1, and EBV-encoded small nonpolyadenylated RNA (EBER)-2 [26].

HSV has been primarily divided into HSV-1 and HSV-2 infection, primarily seen in the oropharyngeal and genital areas. One of the significant features of HSV infection is the virus's ability to remain latent in the sensory ganglia. The role of HSV-1 in the development of OSCC has been evaluated extensively but with contradictory consults. Meurman [27] in a review showed that HSV infection is linked with oral carcinogenesis, but many studies report that HSV-1 and 2 antibodies are not elevated in OSCC pts. Foy et al. [24] proposed that HSV promotes oral carcinogenesis through a hit-and-run mechanism involving epigenome deregulation, which modulates gene expression in response to environmental exposures. Thus a transient infection by Herpesvirus ("hit") can induce epigenetic alterations involving DNA methylation and noncoding RNAs, and these changes may be propagated over many cell divisions even after the clearance of the virus ("run"). Like other oncogenic viruses, Herpesviruses deregulate the essential host genes and pathways, thereby causing carcinogenesis.

7.3 Clinical Implications

Specific bacterial strains can be selectively delivered to tumor sites to mediate the therapeutic responses [5, 28]. The pliability of the microbiome can be directed towards therapeutic approaches via "Vector has driven" antitumor agents. Here, microorganisms can be used as delivery vehicles for protective factors like various proteins, anti-angiogenic factors, and immune therapy. On a futuristic note, marker microorganisms can be tagged, allowing the detection of tumor cells.

References

1. Mascitti M, Togni L, Troiano G et al (2019) Beyond head and neck cancer: the relationship between oral microbiota and tumour development in distant organs. Front Cell Infect Microbiol 9:232
2. Hooper SJ, Wilson MJ, Crean SJ (2009) Exploring the link between microorganisms and oral cancer: a systematic review of the literature. Head Neck 31(9):1228–1239
3. Wang F, Meng W, Wang B, Qiao L (2014) Helicobacter pylori-induced gastric inflammation and gastric cancer. Cancer Lett 345(2):196–202

4. Vimal J, Himal I, Kannan S (2020) Role of microbial dysbiosis in carcinogenesis and cancer therapies. Indian J Med Res 152(6):553–561
5. Cummins J, Tangney M (2013) Bacteria and tumors: causative agents or opportunistic inhabitants? Infect Agent Cancer 8(1):11
6. Al-Hebshi NN, Borgnakke WS, Johnson NW (2019) The microbiome of oral squamous cell carcinomas: a functional perspective. Curr Oral Health Rep 6:145–160
7. Rai AK, Panda M, Das AK et al (2021) Dysbiosis of salivary microbiome and cytokines influence oral squamous cell carcinoma through inflammation. Arch Microbiol 203(1):137–152
8. Chattopadhyay I, Verma M, Panda M (2019) Role of oral microbiome signatures in diagnosis and prognosis of oral cancer. Technol Cancer Res Treat 18:1533033819867354
9. Zhao H, Chu M, Huang Z et al (2017) Variations in oral microbiota associated with oral cancer. Sci Rep 7(1):11773
10. Yang CY, Yeh YM, Yu HY et al (2018) Oral microbiota community dynamics associated with oral squamous cell carcinoma staging. Front Microbiol 9:862
11. Feller L, Altini M, Lemmer J (2013) Inflammation in the context of oral cancer. Oral Oncol 49(9):887–892
12. Karpiński TM (2019) Role of oral microbiota in cancer development. Microorganisms 7(1):20
13. Mazzio EA, Smith B, Soliman KF (2010) Evaluation of endogenous acidic metabolic products associated with carbohydrate metabolism in tumor cells. Cell Biol Toxicol 26(3):177–188
14. Olsen I, Yilmaz Ö (2019) Possible role of *Porphyromonas gingivalis* in orodigestive cancers. J Oral Microbiol 11(1):1563410
15. Lafuente Ibáñez de Mendoza I, Maritxalar Mendia X, GarcÚa de la Fuente AM et al (2020) Role of Porphyromonas gingivalis in oral squamous cell carcinoma development: a systematic review. J Periodontal Res 55(1):13–22
16. Fujiwara N, Kitamura N, Yoshida K et al (2020) Involvement of *Fusobacterium* species in oral cancer progression: a literature review including other types of cancer. Int J Mol Sci 21(17): 6207
17. Patil S (2020) Analyzing the association between *Candida* prevalence, species specificity, and oral squamous cell carcinoma: a systematic review and meta-analysis— Candida and OSCC. Appl Sci 10(3):1099
18. Engku Nasrullah Satiman EAF, Ahmad H, Ramzi AB et al (2020) The role of Candida albicans candidalysin ECE1 gene in oral carcinogenesis. J Oral Pathol Med 49(9):835–841
19. Alnuaimi AD, Ramdzan AN, Wiesenfeld D et al (2016) Candida virulence and ethanol-derived acetaldehyde production in oral cancer and non-cancer subjects. Oral Dis 22(8):805–814
20. Bartie KL, Williams DW, Wilson MJ et al (2004) Differential invasion of Candida albicans isolates in an in vitro model of oral candidosis. Oral Microbiol Immunol 19(5):293–296
21. Krogh P (1990) The role of yeasts in oral cancer by means of endogenous nitrosation. Acta Odontol Scand 48(1):85–88
22. Syrjänen K, Syrjänen S, Lamberg M et al (1983) Morphological and immunohistochemical evidence suggesting human papillomavirus (HPV) involvement in oral squamous cell carcinogenesis. Int J Oral Surg 12(6):418–424
23. Ha PK, Califano JA (2004) The role of human papillomavirus in oral carcinogenesis. Crit Rev Oral Biol Med 15(4):188–196
24. Foy JP, Bertolus C, Boutolleau D et al (2020) Arguments to support a viral origin of oral squamous cell carcinoma in non-smoker and non-drinker patients. Front Oncol 10:822

25. Hillbertz NS, Hirsch JM, Jalouli J et al (2012) Viral and molecular aspects of oral cancer. Anticancer Res 32(10):4201–4212
26. Sivakumar S, Gupta AA, Nik Mohd Rosdy NM et al (2020) Assessing the potential between Epstein-Barr virus and Oral squamous cell carcinoma: a systematic review and meta-analysis. Transl Cancer Res 9(4):3092–3100
27. Meurman JH (2010) Infectious and dietary risk factors of oral cancer. Oral Oncol 46(6): 411–413
28. Baban CK, Cronin M, O'Hanlon D et al (2010) Bacteria as vectors for gene therapy of cancer. Bioeng Bugs 1(6):385–394

Microbial Dysbiosis in Oral Cancer

8

Shaheerah Khan, Rohita Sinha, Saurav Sarkar, Anshuman Dixit, and Samapika Routray

Abstract

Dysbiosis has been proposed to be one of the multiple etiological factors in development and progression of many cancers. Tobacco, alcohol, and betel nut are the common causative factors of oral cancer, which might also lead to the alteration of the oral microbiome composition. Based on knowledge of the carcinogenic effects of dysbiosis, it was recently suggested that probiotics may have antitumor activity. This chapter reviews the role of microbial dysbiosis in oral cancer pathogenesis.

Keywords

Microbial dysbiosis · Eubiosis · *F. nucleatum* · *P. gingivalis* · Oral cancer

S. Khan · A. Dixit
Institute of Life Science, Bhubaneswar, Odisha, India
e-mail: khshaheerah@gmail.com; anshumandixit@gmail.com

R. Sinha
University of Nebraska-Lincoln, Lincoln, NE, USA
e-mail: rohita.sinha@unl.edu

S. Sarkar
Department of ENT, All India Institute of Medical Sciences, Bhubaneswar, Odisha, India
e-mail: ent_saurav@aiimsbhubaneswar.edu.in

S. Routray (✉)
Department of Dentistry, All India Institute of Medical Sciences, Bhubaneswar, Odisha, India
e-mail: dent_samapika@aiimsbhubaneswar.edu.in

S. Routray (ed.), *Microbes and Oral Squamous Cell Carcinoma*,
https://doi.org/10.1007/978-981-19-0592-6_8

8.1 Introduction

The presence of microbes marks the state of that environment. Many microbes maintain the healthy conditions yet there are certain environmental effects (smoking, drinking, poor diet, and health) and few microbes whose presence causes changes in the oral cavity. These changes are observed in gradual dysbiosis followed by inflammation that might finally lead to cancer. It is important to study every molecular detail to understand the role of microbes in the early stages of the disease, which later turns into a cancer. The presence of microbes especially bacteria has been seen to cause many molecular disruptions including the gene, protein, and metabolite level changes. The process of these microbes controlling the host defense system and repair mechanisms and influencing it by taking a hold on cascading pathways and processes of the host is important to understand the alternative causes of cancer other than genetic or epigenetic changes only. The chapter discusses the initial state of cancer like the dysbiosis and the role of microbes in detail, in causing such conditions.

8.2 Oral Microbiome and Human Health

Oral cavity is one of the easiest ways for the microbes to inhabit and get access to the body. The bacteria have been shown to be linked with many oral diseases like periodontitis, dental caries, otitis, tonsillitis, etc. and systemic diseases [1] such as cystic fibrosis, cardiovascular disease [2, 3], rheumatoid arthritis, Alzheimer's disease, diabetes [4], stroke [5], premature birth [6], pneumonia and even cancers [7, 8]. The microbes residing in the oral cavity play important roles in the regulation of many biological pathways and processes through which they are able to alter other systems of the human body as well.

It is not only the bacterial community, but also other microbes together that causes these oral conditions. Bacterial species such as *Porphyromonas gingivalis*, *Treponema denticola*, *Tannerella forsythia*, *Filifactor alocis*, *Parvimonas micra*, and *Aggregatibacter actinomycetemcomitans* and certain archaea like *Methanobrevibacter oralis*, *Methanobacterium curvum*, and *Methanosarcina mazei* are abundantly found in periodontitis [9–14]. Similarly in dental caries microbes that are most likely to be found are the bacterial species *Streptococcus mutans*, and *Lactobacillus sp.* and Fungi *Candida albicans* [15, 16].

The abundance of certain microbes in oral conditions indicates that they may change the structure of the oral cavity, but it is more surprising how these microbes inhabiting the oral cavity is also capable of influencing the other systems of our body. Cystic fibrosis is a genetically acquired disease, where excessive production of mucous blocks the lungs and other organs such as pancreas and the digestive system as a whole. Bacterial species like *Streptococcus oralis*, *Streptococcus mitis*, *Streptococcus gordonii*, and *Streptococcus sanguinis* are known to increase in number in cystic fibrosis [17]. Cardiovascular disease, which is a life-threatening condition, has also been associated with certain bacterial species like *Campylobacter rectus*,

Porphyromonas gingivalis, *Porphyromonas endodontalis*, *Prevotella intermedia*, and *Prevotella nigrescens* [18, 19]. Bacterial influence has also been shown in auto-immune diseases such as rheumatoid arthritis. There was an increase in the number of bacterial species *Rothia mucilaginosa*, *Rothia dentocariosa*, *Lactobacillus salivarius*, and *Cryptobacterium curtum*, in this condition [20, 21]. *Porphyromonas gingivalis* has been shown to be associated with Alzheimer's disease [22–24], while diabetic patients have high number of bacteria from Genera *Aggregatibacter*, *Neisseria*, *Gemella*, *Eikenella*, *Selenomonas*, *Actinomyces*, *Capnocytophaga*, *Fusobacterium*, *Veillonella*, and *Streptococcus* [25]. It has been reported that *H. pylori* may have a protective effect against inflammatory bowel disease (IBD) and that incidence of IBD increased in European population after its eradication [26]. It is evident that the microbiome composition has a direct relation with host physiology; however, it is important to see whether they are the cause or the consequence of a pathological condition.

8.3 Microbes and Human Health

The role of microbes has an important role in predisposing the healthy or unhealthy conditions in the oral cavity. It consist of mainly bacteria, virus, protozoa, fungi, and archaea. Bacterial biofilms help to maintain a healthy environment, yet when disturbed create dysbiosis, leaving the oral cavity in a diseased state [11]. This is followed by chronic inflammatory conditions such as periodontitis [12], making the environment susceptible to cancer. Two specific bacterial species *Fusobacterium nucleatum* and *Porphyromonas gingivalis* play an active role in cancer development, by hijacking and suppressing important processes like cellular repair mechanisms and apoptosis [13, 14], and activating immune evasions, that ultimately leads to carcinogenesis [15]. Viruses such as the human papilloma virus (HPV) are already known from the studies to cause oral, oro-pharyngeal, sino-nasal, and laryngeal carcinomas. The presence of protozoans like *Toxoplasma gondii*, *Cryptosporidium parvum*, and *Trichomonas vaginalis* is known to cause cancer like gliomas, astrocytomas, colorectal cancer, cervical cancer, etc., by causing dysbiosis and blocking the same pathways as other species. It is highly suspected to cause oral cancer as well. The study on fungi especially *Candida* sp. has shown its potency of causing dysbiosis and then cancer. Hence, the presence of these disease-causing microbes is itself a marker for the condition of the oral cavity.

8.4 Microbiome and Oral Cancer

Cancer is a very heterogenous disease, seen mainly to be disrupted on the genetic level and also many epigenetic factors play a role in causing it. Most of the studies involve the analysis of oral cancer in the light of molecular biology yet fail to accommodate the alternative routes through which the host system can be disrupted. The field of the microscopic living organisms that survive, depend on us and influence our behavior, remains unnoticed, which has quite a potential to cause a change that might lead to cancer. There has been studies that indicate the role of these microbes that alter the host genetic system and cause biochemical changes leading to inhibition of protective pathways or activation of inflammatory processes, making the environment the hub of cancer developing cells. It is important to understand the role of these microbes in the process of carcinogenesis, especially in the transition stage that leads to cancer, which is dysbiosis.

8.5 Dysbiosis

A healthy oral microbiome maintains a normal homeostasis, promoting a beneficial mutual, symbiotic, or commensal relationship, without causing any harm to the host, which is referred to as "eubiosis" [27–29]. Though the eubiotic oral microbiome resists any change in its environment, yet due to certain perturbations or changes, the balance shifts into a diseased state, known as "dysbiosis" [30]. Dysbiosis is defined as the overall loss of the normal microbial diversity with a loss of not only the number of beneficial microbes and genetic variability but also gain or increase in the number of pathogenic microbes [31]. Sometimes dysbiosis is also a consequence of acidification of the oral cavity that causes a loss of diversity. Loss of beneficial microbes which are responsible for maintaining the oral immune response is no longer able to protect the environment from the carcinogenic metabolites of pathogenic bacteria, leading to inflammation. Sulfides released by some bacterial genus Bacteroides, nitrosamines by phylum Firmicutes, hydrogen peroxide by the family Enterococcus, and deoxycholic acid (DCA) by *Clostridium scindens* are few metabolites that lead to carcinogenesis, by degrading the mucins (a protective barrier), releasing free radicals that cause DNA breaks and cross-linking, etc. They are made through the biochemical processes by some bacteria that ferment the host protein into these metabolites, which can also induce mutations in the host [32]. The inflammation results into conditions like periodontitis and dental caries, making the host bloodstream easily accessible to the pathogens. This helps the bacteria move from oral cavity to other tissues through systemic circulation. A continuous degrading situation like this leads to an increase in the number of carcinogenic bacteria in the oral cavity, which is potent enough to change the dysbiotic oral tissues into a cancerous state [33]. A dysbiotic oral microbiome has been found having its own microbial signature which mostly includes *Treponema denticola*, *Tannerella forsythia*, *Porphyromonas gingivalis*, *Fusobacterium nucleatum*, *Prevotella intermedia*, and *Parvimonas micra* [34, 35]. These bacteria are able to control cell proliferation by activating the nuclear factor κB (NF-κB) that promotes carcinogenesis [36–38]. The interaction among the microbes causes dysbiosis, breaking the epithelial barriers and causing inflammation which is potent enough to predispose neoplastic transformation in the oral cavity, leading to the oral squamous cell carcinoma [39–43] (Fig. 8.1).

The mechanism through which these bacteria are able to enter and cause changes in the environment has been studied extensively. A study highlighted the concept of

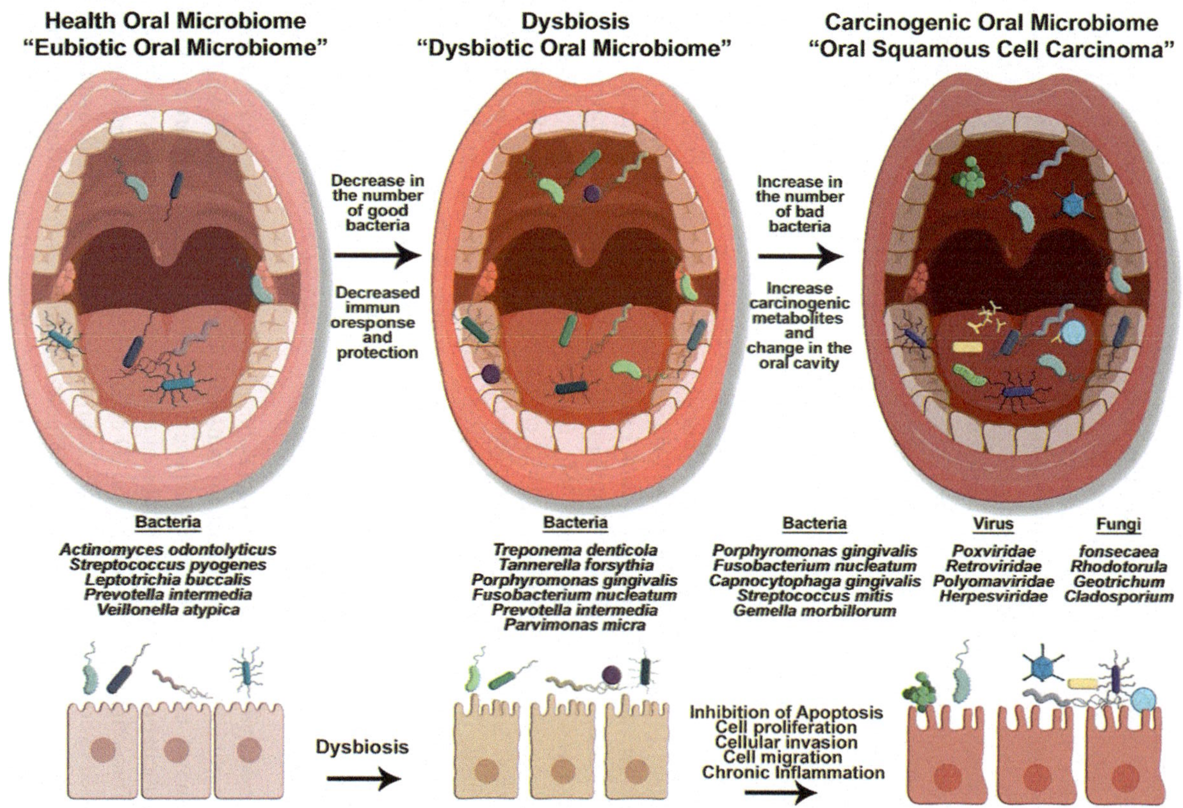

Fig. 8.1 The transition of a healthy human oral cavity into a dysbiotic and then carcinogenic environment. The decrease in the number of good microbes and increase in the number of bad microbes compromise the system by decreasing its anti-inflammatory response and protection, and increasing the carcinogenic metabolites that cause a change in the oral cavity. A dysbiosis, finally, leads to chronic inflammation aiding the OSCC

tumor passenger microbiome and tumor driver microbiome. The concept comes from the "passenger-driver" model in genomics. Just like the driver genes in the body can drive the process of cancer, similarly certain microbes are also capable of driving the process of cancer. A normal oral cavity contains certain community of passenger microbes, which also harbor a subset of species with certain competitive advantage, like genera *Fusobacterium*, *Prevotella*, *Campylobacter*, *Selenomonas*, and *Candida albicans* that express certain pro-inflammatory factors like LPS biosynthesis, flagellar assembly, chemotaxis, etc. changing it into a dysbiotic "driver" environment that can help drive the process of cancer. Many studies have pointed towards their mechanism of action where it starts from the production of carcinogenic metabolites [44], such as acetaldehyde, nitrite, hydrogen sulfide, and short chain fatty acids [45, 46], followed by the inhibition of apoptosis, activation of cell proliferation and invasion, and producing chronic inflammation. They activate the beta-catenin and metalloprotease nine pathways that lead to carcinogenesis through virulence factors FadA and gingipain [47].

Studies have shown how the bacteria aids to cause inflammation with their metabolic by-products and by changing the pH of the oral cavity, which can be the premalignant conditions that lead to cancer. Studies have shown the presence and increase in the number of *Fusobacterium nucleatum* in colorectal cancer [48]; *Porphyromonas gingivalis* and *Aggregatibacter actinomycetemcomitans* in pancreatic cancer [49, 50]; *Helicobacter pylori* in gastric cancer [51]; *Salmonella typhi* in gallbladder cancer [52]; and *Bacteroides fragilis* in colon cancer [53]. Comparative studies between the tumor and nontumor tissues taken from the oral cavity showed an increase in the number of *Capnocytophaga gingivalis*, *Prevotella melaninogenica*, *Streptococcus mitis*, *Peptostreptococcus stomatis*, *Streptococcus salivarius*, *Streptococcus gordonii*, *Gemella haemolysans*, *Gemella morbillorum*, *Johnsonella ignava*, and *Streptococcus parasanguinis* in oral cancer [54–56], while esophageal cancer showed an increase in the number of *Tannerella forsythia*, and *Porphyromonas gingivalis* [57]. However, *Porphyromonas gingivalis* and *Fusobacterium nucleatum* have shown to be quiet famous in oral cavity cancer. *Porphyromonas gingivalis* (*P. gingivalis*) and *Fusobacterium nucleatum* (*F. nucleatum*) have been associated with a wide variety of cancers including oral cancer.

Porphyromonas gingivalis: Studies have shown the presence of *P. gingivalis* abundantly in oral samples of patients suffering from gingival diseases, while others have shown it as a contributor to the progression of oral cancer. They are involved in the epithelial-mesenchymal transition of malignant cells, proliferation, and invasion [58]. Transcriptomic study of the Cal-27 cells infected with *P. gingivalis* showed an increased expression of *CD47*, *IL-1R2 suprabasin*, and *IL-1α*, *CCL-3*, and *CCL-5* in macrophages, which encodes pro-tumor molecules. It helps in invasion by inhibiting the phagocytosis of Cal-27 cells by macrophages. Mice infected with *P. gingivalis* not only induce polarization of macrophages into M2 tumor-associated macrophages, but also promoted the growth of oral squamous cell carcinoma [13]. *P. gingivalis* is associated with the TNM (T: primary tumor, N: regional lymph nodes, M: metastases) cancer staging, such that they were found higher in patients with TNM stage 3 and 4 (late stages), with poorly differentiated tissue and lymph node metastasis. It overexpresses matrix metalloproteinases, which degrade the extracellular matrix, promoting cell invasion and metastasis [59]. *P. gingivalis* suppresses the activity of p53, increases the level of Cyclin A, and activates PI3K-AKT pathway promoting cell proliferation [60]. They have also shown an increase in alpha-defensin, promoting oral squamous cell carcinoma (OSCC) [61].

Fusobacterium nucleatum: A study investigated the effects of the periodontal bacteria in oral cancer, to check the effects of the bacteria on the cells, the immunological changes, and the oncogenic expression change. *F. nucleatum* comparatively showed greater effects on the cancer cells. Immunological changes included an increase in the expression of IL8. IL8 or CXCL8 promotes angiogenesis [62], stimulates chemotaxis of neutrophils and basophils [63], and it enhances tumor growth and invasion in cancer cell lines [64, 65]. It has been shown that blocking of CXCL8 selective antagonists inhibits CXCL8-mediated tumor growth and invasion in cancer models [66, 67]. It also plays an important role in cancer stem cell renewal and survival, especially in breast cancer; it has been seen to enhance the self-renewal of ALDH1+ cells and promote breast cancer cell survival through CXCR1-dependent mechanisms [66]. CXCL8 stimulated growth of tumor spheroids in nasopharyngeal carcinoma, through a PI-3K/AKT- and CXCR2-dependent mechanism [68]. The presence of *F. nucleatum* also increases the expression of cell survival markers *MYC*, *JAK1,* and *STAT3* and epithelial-mesenchymal transition markers *ZEB1* and *TGF-β* [69]. *F. nucleatum* has been seen to let its protein Fap2 interact with inhibitory immune-receptor TIGIT on NK and T cells, which protects tumors from immune cell attack [9]. *F. nucleatum* activates kinases to induce proliferation of oral epithelial cell. The FadA protein of *F. nucleatum* binds to E-cadherin and activates beta-catenin signaling regulating cell proliferation and inflammatory responses [11, 12, 70]. *F. nucleatum* also works synergistically with other bacteria, fungi, and viruses to induce pro-inflammatory microenvironment by recruiting tumor-infiltrating immune cells that cause tumorigenesis [15].

Through these mechanisms *Fusobacterium nucleatum* and *Porphyromonas gingivalis* are able to increase selectivity and multiplicity promoting tumor promotion and invasion. In mice models it has been seen to express certain pro-inflammatory signatures which are correlated with those present in human colorectal cancers. *F. nucleatum* recruits tumor-infiltrating immune cells, to generate a pro-inflammatory microenvironment such that it helps in the progression of cancer [71] (Fig. 8.2).

8.6 Conclusion

The changing composition of oral microbiome in oral diseases indicates there is a major bridging, where either the microbes influence the cause or the disease condition alters their number, which might again synergistically increase the spread of the disease. Either way the spotlight falls on the microbes of the oral cavity. This leads to the question of the influence of the oral microbes on oral cancer.

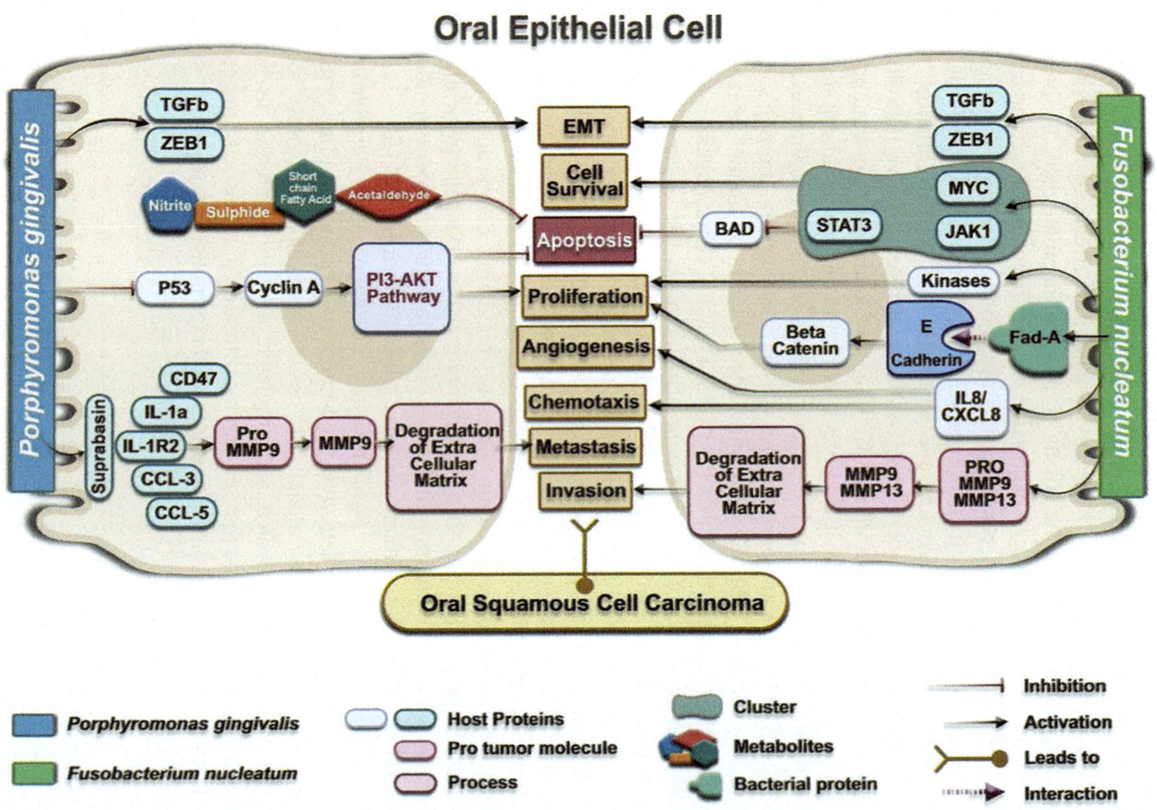

Fig. 8.2 The molecular mechanism involved in the progression of cancer, contributed by the presence of *Porphyromonas gingivalis* and *Fusobacterium nucleatum*, in the oral epithelial cells of the human oral cavity

References

1. Seymour GJ, Ford PJ, Cullinan MP, Leishman S, Yamazaki K (2007) Relationship between periodontal infections and systemic disease. Clin Microbiol Infect 13(Suppl 4):3–10
2. Beck JD, Offenbacher S (2005) Systemic effects of periodontitis: epidemiology of periodontal disease and cardiovascular disease. J Periodontol 76(Suppl 11S):2089–2100
3. Joshipura KJ, Rimm EB, Douglass CW, Trichopoulos D, Ascherio A, Willett WC (1996) Poor oral health and coronary heart disease. J Dent Res 75(9):1631–1636
4. Genco RJ, Grossi SG, Ho A, Nishimura F, Murayama Y (2005) A proposed model linking inflammation to obesity, diabetes, and periodontal infections. J Periodontol 76(11 Suppl):2075–2084
5. Joshipura KJ, Hung HC, Rimm EB, Willett WC, Ascherio A (2003) Periodontal disease, tooth loss, and incidence of ischemic stroke. Stroke 34(1):47–52
6. Offenbacher S, Jared HL, O'Reilly PG, Wells SR, Salvi GE, Lawrence HP et al (1998) Potential pathogenic mechanisms of periodontitis associated pregnancy complications. Ann Periodontol 3(1):233–250
7. Aas JA, Paster BJ, Stokes LN, Olsen I, Dewhirst FE (2005) Defining the normal bacterial flora of the oral cavity. J Clin Microbiol 43(11):5721–5732
8. Duran-Pinedo AE, Frias-Lopez J (2015) Beyond microbial community composition: functional activities of the oral microbiome in health and disease. Microbes Infect 17(7):505–516
9. Costalonga M, Herzberg MC (2014) The oral microbiome and the immunobiology of periodontal disease and caries. Immunol Lett 162(2 Pt A):22–38
10. Griffen AL, Beall CJ, Campbell JH, Firestone ND, Kumar PS, Yang ZK et al (2012) Distinct and complex bacterial profiles in human periodontitis and health revealed by 16S pyrosequencing. ISME J 6(6):1176–1185
11. Haubek D (2010) The highly leukotoxic JP2 clone of Aggregatibacter actinomycetemcomitans: evolutionary aspects, epidemiology and etiological role in aggressive periodontitis. APMIS Suppl (130):1–53
12. Jorth P, Turner KH, Gumus P, Nizam N, Buduneli N, Whiteley M (2014) Metatranscriptomics of the human oral microbiome during health and disease. MBio 5(2):e01012–e01014
13. Liu S, Zhou X, Peng X, Li M, Ren B, Cheng G et al (2020) Porphyromonas gingivalis promotes immunoevasion of oral cancer by protecting cancer from macrophage attack. J Immunol 205(1): 282–289
14. Vartoukian SR, Palmer RM, Wade WG (2009) Diversity and morphology of members of the phylum "synergistetes" in periodontal health and disease. Appl Environ Microbiol 75(11): 3777–3786
15. Koo H, Bowen WH (2014) Candida albicans and Streptococcus mutans: a potential synergistic alliance to cause virulent tooth decay in children. Future Microbiol 9(12):1295–1297
16. Willis JR, Gabaldon T (2020) The human oral microbiome in health and disease: from sequences to ecosystems. Microorganisms 8(2)
17. Whiley RA, Fleming EV, Makhija R, Waite RD (2015) Environment and colonisation sequence are key parameters driving cooperation and competition between Pseudomonas aeruginosa cystic fibrosis strains and oral commensal streptococci. PLoS One 10(2):e0115513
18. Chhibber-Goel J, Singhal V, Bhowmik D, Vivek R, Parakh N, Bhargava B et al (2016) Linkages between oral commensal bacteria and atherosclerotic plaques in coronary artery disease patients. NPJ Biofilm Microbiom 2:7
19. Teles R, Wang CY (2011) Mechanisms involved in the association between periodontal diseases and cardiovascular disease. Oral Dis 17(5):450–461
20. Roszyk E, Puszczewicz M (2017) Role of human microbiome and selected bacterial infections in the pathogenesis of rheumatoid arthritis. Reumatologia 55(5):242–250
21. Scher JU, Ubeda C, Equinda M, Khanin R, Buischi Y, Viale A et al (2012) Periodontal disease and the oral microbiota in new-onset rheumatoid arthritis. Arthritis Rheum 64(10):3083–3094

22. Aguayo S, Schuh C, Vicente B, Aguayo LG (2018) Association between Alzheimer's disease and oral and gut microbiota: are pore forming proteins the missing link? J Alzheimer's Dis 65(1):29–46
23. Dominy SS, Lynch C, Ermini F, Benedyk M, Marczyk A, Konradi A et al (2019) Porphyromonas gingivalis in Alzheimer's disease brains: evidence for disease causation and treatment with small-molecule inhibitors. Sci Adv 5(1):eaau3333
24. Miklossy J (2016) Bacterial amyloid and DNA are important constituents of senile plaques: further evidence of the spirochetal and biofilm nature of senile plaques. J Alzheimer's Dis 53(4): 1459–1473
25. Casarin RC, Barbagallo A, Meulman T, Santos VR, Sallum EA, Nociti FH et al (2013) Subgingival biodiversity in subjects with uncontrolled type-2 diabetes and chronic periodontitis. J Periodontal Res 48(1):30–36
26. Imawana RA, Smith DR, Goodson ML (2020) The relationship between inflammatory bowel disease and Helicobacter pylori across East Asian, European and Mediterranean countries: a meta-analysis. Ann Gastroenterol 33(5):485–494
27. Avila M, Ojcius DM, Yilmaz O (2009) The oral microbiota: living with a permanent guest. DNA Cell Biol 28(8):405–411
28. Dovrolis N, Kolios G, Spyrou GM, Maroulakou I (2019) Computational profiling of the gut-brain axis: microflora dysbiosis insights to neurological disorders. Brief Bioinform 20(3): 825–841
29. Zhang WL, Wang SS, Wang HF, Tang YJ, Tang YL, Liang XH (2019) Who is who in oral cancer? Exp Cell Res 384(2):111634
30. Radaic A, Kapila YL (2021) The oralome and its dysbiosis: new insights into oral microbiome-host interactions. Comput Struct Biotechnol J 19:1335–1360
31. Petersen C, Round JL (2014) Defining dysbiosis and its influence on host immunity and disease. Cell Microbiol 16(7):1024–1033
32. Pang X, Tang YJ, Ren XH, Chen QM, Tang YL, Liang XH (2018) Microbiota, epithelium, inflammation, and TGF-beta signaling: an intricate interaction in oncogenesis. Front Microbiol 9:1353
33. Gomar-Vercher S, Cabrera-Rubio R, Mira A, Montiel-Company JM, Almerich-Silla JM (2014) Relationship of children's salivary microbiota with their caries status: a pyrosequencing study. Clin Oral Investig 18(9):2087–2094
34. Lamont RJ, Koo H, Hajishengallis G (2018) The oral microbiota: dynamic communities and host interactions. Nat Rev Microbiol 16(12):745–759
35. Tonoyan L, Vincent-Bugnas S, Olivieri CV, Doglio A (2019) New viral facets in oral diseases: the EBV paradox. Int J Mol Sci 20(23):5861
36. Dolcet X, Llobet D, Pallares J, Matias-Guiu X (2005) NF-kB in development and progression of human cancer. Virchows Arch 446(5):475–482
37. Guarneri C, Bevelacqua V, Polesel J, Falzone L, Cannavo PS, Spandidos DA et al (2017) NFkappaB inhibition is associated with OPN/MMP9 downregulation in cutaneous melanoma. Oncol Rep 37(2):737–746
38. Leonardi GC, Falzone L, Salemi R, Zanghi A, Spandidos DA, McCubrey JA et al (2018) Cutaneous melanoma: from pathogenesis to therapy (review). Int J Oncol 52(4):1071–1080
39. Couturier-Maillard A, Secher T, Rehman A, Normand S, De Arcangelis A, Haesler R et al (2013) NOD2-mediated dysbiosis predisposes mice to transmissible colitis and colorectal cancer. J Clin Invest 123(2):700–711
40. Francescone R, Hou V, Grivennikov SI (2014) Microbiome, inflammation, and cancer. Cancer J 20(3):181–189
41. Hu B, Elinav E, Huber S, Strowig T, Hao L, Hafemann A et al (2013) Microbiota-induced activation of epithelial IL-6 signaling links inflammasome-driven inflammation with transmissible cancer. Proc Natl Acad Sci U S A 110(24):9862–9867
42. La Rosa GRM, Gattuso G, Pedulla E, Rapisarda E, Nicolosi D, Salmeri M (2020) Association of oral dysbiosis with oral cancer development. Oncol Lett 19(4):3045–3058

43. Ojesina AI, Lichtenstein L, Freeman SS, Pedamallu CS, Imaz-Rosshandler I, Pugh TJ et al (2014) Landscape of genomic alterations in cervical carcinomas. Nature 506(7488):371–375
44. Sharon G, Garg N, Debelius J, Knight R, Dorrestein PC, Mazmanian SK (2014) Specialized metabolites from the microbiome in health and disease. Cell Metab 20(5):719–730
45. Basic A, Blomqvist S, Carlen A, Dahlen G (2015) Estimation of bacterial hydrogen sulfide production in vitro. J Oral Microbiol 7:28166
46. Hezel MP, Weitzberg E (2015) The oral microbiome and nitric oxide homoeostasis. Oral Dis 21(1):7–16
47. Whitmore SE, Lamont RJ (2014) Oral bacteria and cancer. PLoS Pathog 10(3):e1003933
48. Kato I, Vasquez AA, Moyerbrailean G, Land S, Sun J, Lin HS et al (2016) Oral microbiome and history of smoking and colorectal cancer. J Epidemiol Res 2(2):92–101
49. Fan X, Alekseyenko AV, Wu J, Peters BA, Jacobs EJ, Gapstur SM et al (2018) Human oral microbiome and prospective risk for pancreatic cancer: a population-based nested case-control study. Gut 67(1):120–127
50. Torres PJ, Fletcher EM, Gibbons SM, Bouvet M, Doran KS, Kelley ST (2015) Characterization of the salivary microbiome in patients with pancreatic cancer. PeerJ 3:e1373
51. Kim SS, Ruiz VE, Carroll JD, Moss SF (2011) Helicobacter pylori in the pathogenesis of gastric cancer and gastric lymphoma. Cancer Lett 305(2):228–238
52. Nagaraja V, Eslick GD (2014) Systematic review with meta-analysis: the relationship between chronic Salmonella typhi carrier status and gall-bladder cancer. Aliment Pharmacol Ther 39(8): 745–750
53. Toprak NU, Yagci A, Gulluoglu BM, Akin ML, Demirkalem P, Celenk T et al (2006) A possible role of Bacteroides fragilis enterotoxin in the aetiology of colorectal cancer. Clin Microbiol Infect 12(8):782–786
54. Mager DL, Haffajee AD, Devlin PM, Norris CM, Posner MR, Goodson JM (2005) The salivary microbiota as a diagnostic indicator of oral cancer: a descriptive, non-randomized study of cancer-free and oral squamous cell carcinoma subjects. J Transl Med 3:27
55. Pushalkar S, Ji X, Li Y, Estilo C, Yegnanarayana R, Singh B et al (2012) Comparison of oral microbiota in tumor and non-tumor tissues of patients with oral squamous cell carcinoma. BMC Microbiol 12:144
56. Wang L, Ganly I (2014) The oral microbiome and oral cancer. Clin Lab Med 34(4):711–719
57. Peters BA, Wu J, Pei Z, Yang L, Purdue MP, Freedman ND et al (2017) Oral microbiome composition reflects prospective risk for esophageal cancers. Cancer Res 77(23):6777–6787
58. Lafuente Ibanez de Mendoza I, Maritxalar Mendia X, Garcia de la Fuente AM, Quindos Andres G, Aguirre Urizar JM (2020) Role of Porphyromonas gingivalis in oral squamous cell carcinoma development: a systematic review. J Periodontal Res 55(1):13–22
59. Chang C, Geng F, Shi X, Li Y, Zhang X, Zhao X et al (2019) The prevalence rate of periodontal pathogens and its association with oral squamous cell carcinoma. Appl Microbiol Biotechnol 103(3):1393–1404
60. Kuboniwa M, Hasegawa Y, Mao S, Shizukuishi S, Amano A, Lamont RJ et al (2008) P. gingivalis accelerates gingival epithelial cell progression through the cell cycle. Microbes Infect 10(2):122–128
61. Hoppe T, Kraus D, Novak N, Probstmeier R, Frentzen M, Wenghoefer M et al (2016) Oral pathogens change proliferation properties of oral tumor cells by affecting gene expression of human defensins. Tumour Biol 37(10):13789–13798
62. Koch AE, Polverini PJ, Kunkel SL, Harlow LA, DiPietro LA, Elner VM et al (1992) Interleukin-8 as a macrophage-derived mediator of angiogenesis. Science 258(5089): 1798–1801
63. Geginat J, Larghi P, Paroni M, Nizzoli G, Penatti A, Pagani M et al (2016) The light and the dark sides of Interleukin-10 in immune-mediated diseases and cancer. Cytokine Growth Factor Rev 30:87–93

64. Schadendorf D, Moller A, Algermissen B, Worm M, Sticherling M, Czarnetzki BM (1993) IL-8 produced by human malignant melanoma cells in vitro is an essential autocrine growth factor. J Immunol 151(5):2667–2675
65. Singh RK, Gutman M, Reich R, Bar-Eli M (1995) Ultraviolet B irradiation promotes tumorigenic and metastatic properties in primary cutaneous melanoma via induction of interleukin 8. Cancer Res 55(16):3669–3674
66. Ginestier C, Liu S, Diebel ME, Korkaya H, Luo M, Brown M et al (2010) CXCR1 blockade selectively targets human breast cancer stem cells in vitro and in xenografts. J Clin Invest 120(2):485–497
67. Khan MN, Wang B, Wei J, Zhang Y, Li Q, Luan X et al (2015) CXCR1/2 antagonism with CXCL8/Interleukin-8 analogue CXCL8(3-72)K11R/G31P restricts lung cancer growth by inhibiting tumor cell proliferation and suppressing angiogenesis. Oncotarget 6(25): 21315–21327
68. Lo MC, Yip TC, Ngan KC, Cheng WW, Law CK, Chan PS et al (2013) Role of MIF/CXCL8/CXCR2 signaling in the growth of nasopharyngeal carcinoma tumor spheres. Cancer Lett 335(1):81–92
69. Harrandah AM, Chukkapalli SS, Bhattacharyya I, Progulske-Fox A, Chan EKL (2020) Fusobacteria modulate oral carcinogenesis and promote cancer progression. J Oral Microbiol 13(1):1849493
70. Liu B, Faller LL, Klitgord N, Mazumdar V, Ghodsi M, Sommer DD et al (2012) Deep sequencing of the oral microbiome reveals signatures of periodontal disease. PLoS One 7(6): e37919
71. Kostic AD, Chun E, Robertson L, Glickman JN, Gallini CA, Michaud M et al (2013) Fusobacterium nucleatum potentiates intestinal tumorigenesis and modulates the tumor-immune microenvironment. Cell Host Microbe 14(2):207–215

Bacteria-Induced Carcinopathogenesis in Oral Squamous Cell Carcinoma

9

Swagatika Panda, Divya Gopinath, and Newell W. Johnson

Abstract

The human oral microbiome plays an important role in host physiology. The causes and factors that influence the microbiota are closely linked to the cause of a disease. Some oral taxa, especially *Porphyromonas gingivalis* and *Fusobacterium nucleatum*, induce carcinogenesis by several different mechanisms, like inhibiting apoptosis, inducing cell proliferation, cellular invasion, migration and chronic inflammation. These changes are seen parallel with the malignant lesions developed in the oral cavity. Starting with a brief introduction on eubiosis and dysbiosis of oral microbiome this chapter has highlighted the role of structural and metabolic attributes of bacteria in oral carcinogenesis via influencing the eight established hallmarks of cancer.

Keywords

Bacteria · Carcinopathogenesis · OSCC

S. Panda (✉)
Department of Oral Pathology and Microbiology, Institute of Dental Sciences, Siksha 'O' Anusandhan Deemed to be University, Bhubaneswar, Odisha, India
e-mail: swagatikapanda@soa.ac.in

D. Gopinath
Clinical Oral Health Sciences Division, School of Dentistry, International Medical University, Kuala Lumpur, Malaysia
e-mail: divyagopinath@imu.edu.my

N. W. Johnson
Menzies Health Institute Queensland and School of Medicine and Dentistry, Griffith University, Brisbane, QLD, Australia

Faculty of Dentistry, Oral and Craniofacial Sciences, King's College London, London, UK
e-mail: n.johnson@griffith.edu.au; newell.johnson@kcl.ac.uk

S. Routray (ed.), *Microbes and Oral Squamous Cell Carcinoma*,
https://doi.org/10.1007/978-981-19-0592-6_9

9.1 Introduction

Head and neck cancer encompasses a complex group of cancers arising from the mucous membranes of the upper aero-digestive tract; skin, nasopharynx, nasal mucosa and associated air cells, oropharynx, oral cavity, larynx, hypopharynx and salivary glands. In discussing cancers of these many sites, it is always recommended to specify the nature and site of the cancer by use of discrete codes of the International Classification of Diseases (ICD), the eleventh iteration of which is due to come into effect on January 1, 2022 [1]. Detailed histological subtypes are coded by ICD for Oncology, ICD-O, and the latest iteration being recommended for use from 2020 [2]. Head and neck malignancies are significant causes of morbidity and mortality, particularly in resource-limited countries [3, 4]. More than 90% of oral cancers are oral squamous cell carcinoma (OSCC) which is the focus in this chapter. Further, different anatomical subsites in OSCC need to be studied separately as they differ in their aetiologies and behaviours. Several risk factors have been correlated with the development of oral cancer among which tobacco, alcohol and bad oral hygiene are the most studied [5–7]. Although microorganisms have long been speculated to be involved in carcinogenesis, the understanding of human microbial interaction with human cancers has evolved only over recent decades chiefly attributed to modern sequencing technologies. Chronic infection with Epstein-Barr virus in nasopharyngeal carcinomas, human papilloma virus and periodontopathic bacteria in oral and oropharyngeal cancers are the much-discussed microorganisms. Interest in the potential role of bacteria and oncogenesis has been on a rise since the classification of *Helicobacter pylori* as a class 1 carcinogen by the World Health Organization [8]. This chapter summarises and critiques the evidence supporting a role for bacteria in oral carcinogenesis.

9.2 Oral Bacteriome and Eubiosis

The Greek term 'Eubiosis' refers to an optimal balance of microflora at a particular anatomic site, originally used in relation to the gastrointestinal tract [9]. The human microbiome was initially defined as an ecological community of symbiotic microorganisms [10] which was later modified as the collective genome of the microbes that live in and on the human body [11, 12]. Bacteria, viruses, fungi, algae and small protists are members of microbiomes which interact with each other and form their ecological niche together. Later the bacterial community was termed as 'bacteriome'. The microbiome influences the host through the metabolic products of the organisms present which is referred to as metabolome [13]. The overall microbial metabolome is a complex interaction with the host [11]. Our whole body contains at least as many bacteria as human cells (est. 3.0×10^{13} human and est. 3.8×10^{13} microbial cells) [14]. In this chapter we concentrate on the bacteriome and bacteriomics, whilst being cognisant of the dangers of ignoring the role of other organisms. Fortunately, as whole genome sequencing becomes more widely available and affordable, information on comprehensive metabolomics of the mouth in health and a range of diseases is becoming available.

The oral cavity is known to harbour the second most diverse bacteriome in the human body, harbouring over 700 bacterial species [15] which are included in 185 genera and 12 phyla. The phyla include Fusobacteria, Proteobacteria, Firmicutes, Actinobacteria, Bacteroidetes, Chloroflexi, Chlamydiae, Spirochaetes, Synergistetes, Saccharibacteria, Gracilibacteria and Absconditabacteria [16]. The core microbiome is thought to be common to all humans, whereas a part of the variable microbiome is unique to each human according to physiological differences [17]. The oral cavity in health typically presents numerous niches for bacteria to colonise where they grow as interactive communities of structurally and functionally organised biofilms unique to their ecological niche, with its distinctive microenvironmental conditions [18]. These niches are highly heterogeneous and each evolves to encompass, in health, the results of continuous interchanges between the nutrients available and the sharing of metabolic products between members of the community [19–21]. Supragingival plaque biofilms usually harbour bacterial genera that can be correlated with maintaining health of the periodontium whereas subgingival niches are inhabited by microorganisms possessing the potential for production of metabolites inimical to periodontal health [20, 21]. Teeth have non-shedding surfaces, permitting the development of biofilms which mature over time to become very different from those in the early days of formation: mature dental plaque can evolve, depending on diet and the degree of disturbance from oral hygiene measures to be acidogenic and cariogenic, or alkaline thus forming dental calculus [21]. Dental restorations, removable prostheses and implants can also function as additional non-shedding surfaces that can influence persistence of bacteria within biofilms [22, 23]. The tongue and other mucosal surfaces harbour dense microbial communities [24, 25]. The oral bacteriome is influenced by age, dentition, ethnicity [26, 27], diet [28, 29], habits [30, 31] and hygiene practices [32–34]. Development and maturation of both innate and adaptive mucosal immunity have been demonstrated to be dependent upon microbiome-based immune education [35]. Earliest evidence of this interplay between microbes and the immune system was described as early as the 1950s [36]. A recent review [35] narrated the mechanisms of gut microbiota enhancing myelopoiesis, maturation of myeloid and Kupffer cells, induction of innate lymphoid cells and influencing Th9 development.

A bacteriome is considered to be eubiotic when it involves a preponderance of potentially beneficial bacteria, with potentially pathogenic species present in low concentrations so that the homeostatic balance of the particular ecosystem is maintained: microbe-host eubiotic interactions are keys in maintaining the homeostasis of all our body systems.

9.3 Microbial Dysbiosis

Dysbiosis is defined as change in the constitution and role of the commensal microorganisms which are observed in healthy human beings [37–39]. Peterson and Round [37] have described microbial dysbiosis into three mutually non-exclusive concurrent events. Loss of beneficial microbes is the first step of dysbiosis causing subsequent expansion of harmful commensals which are known as pathobionts. This leads to inflammation and diseases. Thus the diversity of the microbiota is lost which impacts disease progression. In the past decade, there have been many studies investigating a link between commensal microbiota and cancers [38, 40, 41]. Bacteria can have direct and indirect effects on the tumour microenvironment which consists of transformed, malignant, proliferative and apoptotic keratinocytes, fibro-vascular stroma and an immune inflammatory response. Bacteria generate toxins that can damage DNA [42, 43], boost the proliferation potential of many cell types [44], have immunomodulatory effects [35, 44], can alter mucosal permeability [44, 45], and induce the development of invasive phenotypes which enable metastases [45, 46].

To date, the majority of the evidence on the role of dysbiosis in carcinogenesis is resultant of in vitro experiments which fail to completely reproduce the role of microbiota in the convoluted human health system. The evolving next-generation sequencing technologies have started offering novel interpretations of the role of microbes in health and disease through facilitating the understanding of uninterrupted interaction between the human cells and innumerable commensal and non-commensal bacterial strains of the human body. This advanced technology reveals not only the pathogenesis of disease caused by an individual microorganism but also the entire structural and functional aspects of human microbiota. There is cumulative recognition of the importance of polymicrobial interactions in human diseases. Hence meta-genomic equivalents of Koch postulates applied to non-infectious microbiome-associated diseases have been proposed which are enumerated below [47].

- A single strain, species, genus or phylum, or various combinations thereof, or genomically encoded functions, are significantly correlated with the disease phenotype of a host.
- Microbiota transfer into an appropriate gnotobiotic defined animal model causes at least some aspects of the disease phenotype or a corresponding change of the metabolome or immune response.
- The qualitative or quantitative metagenome alteration should be observed in the newly diseased host.
- Depletion of the identified taxa or functions by intervention via, e.g. antibiotics, expansion of beneficial microbes by probiotics, prebiotics, microbiota transfer or diet reduces progression or ameliorates the disease.
- Transmissibility is not essentially required as diseases are often the result of a shift of the abundance of endogenous members/functional capacities of the microbiota.

These may help to provide a conceptual framework for experimental validation of the hypotheses inferred from genomic analysis of microbiomes.

9.3.1 Dysbiosis During Oral Carcinogenesis

Imbalances in bacterial abundances have been associated with many human cancers and current evidence indicates a role in oncogenesis [48]. Using culture methods, Nagy et al. illustrated differences in the composition of the microbial communities in OSCC and adjacent normal tissue [49]. One of the earliest reports detected *Streptococcus anginosus (S. anginosus)* in OSCC tissues through Polymerase Chain Reaction (PCR) [50, 51], which was later recognised as a normal coloniser of the human oral cavity [52]. Hooper et al. in 2006 were the first to demonstrate the existence of a viable bacterial community within OSCC tissues using both culture and speciation by 16S r RNA sequencing [53]. Gopinath et al. in 2019 [40] emphasised the necessity of a metagenomic approach to understand the whole microbiome and metabolome as well. To date, no single bacterium could be implicated in the aetiology of OSCC. The most common bacteria reported to be frequently associated with OSCC are gram-negative and anaerobic periodontopathic bacteria [40]. Among these, the most common genera repeatedly found enriched in the mouths of OSCC patients, identified using diverse molecular approaches, are of the genus *Fusobacteria.* Various members of *Fusobacteria* are isolated directly from fresh biopsy tissues [54], saliva [55] and surface swabs [56] across different populations [57], including North American [58], Taiwanese [55] and Chinese cohorts [56]. An interesting observation regarding *Fusobacteria* was reported by Yang et al. in 2018 who noticed an increased abundance of bacteria from this genus in oral rinses of patients with stage 4 OSCC compared to stage 1 [55]. The species *Porphyromonas gingivalis* (*P. gingivalis*) have been reported several times in OSCC [59]. Gingival carcinoma has shown a 33% increased prevalence of *P. gingivalis* compared to the gingiva [60]. *P. gingivalis* mediated augmented proliferation rate of human oral cancer cells in vitro [61, 62] and in 4-Nitroquinoline-1-Oxide-induced tongue tumours of mice [63]. *P gingivalis* was shown to increase the tumour proliferation individually and in combination with *Fusobacterium* [64]. Other bacteria found in increased amounts in oral cancer tissues include *Veillonella* [65], *Lactobacillus* [66], *Prevotella* [67], *Staphylococcus* [68], *Lautropia* [54], *Peptostreptococcus* [69], *E. coli* [69], *Capnocytophaga* [54] and *Pseudomonas* [70, 71].

Given the ease and non-invasive nature of sample collection, saliva was studied in several researches to explore the associations of bacteria with oral cancer. Genera reported to be elevated in saliva of OSCC patients included *Parvimonas* [54], *Haemophilus* [67, 69], *Prevotella* [67, 72], *Campylobacter* [67], *Capnocytophaga* [72], *Peptostreptococcus* [67], *Filifactor* [67], *Neisseria* [73], *Oribacterium* [70], *Proteobacteria* [74] and *Actinobacteria* [74, 75]. The presence of these bacteria in tumour niches may well contribute to tumour progression by common traits such as via pro-inflammatory pathways.

9.3.2 Mechanistic Role of Bacteria in Oral Carcinogenesis

The host immune system can be modulated by bacteria on the surface of oral mucosa as well as by bacteria which have actually invaded the tumours. Structural attributes and metabolic products of many bacteria may initiate and promote oral carcinogenesis via influences on the established hallmarks of cancer: tumour-promoting inflammation and genomic stability in particular. Bacteria initiate chemotaxis of immune cells and generate dynamic interactions between microbe-associated molecular pattern (MAMP) and Pattern Recognition Receptors (PRR) of the immune cells to release chemokines and cytokines, thus making modulation of bacteria at remote sites possible.

9.3.2.1 Tumour-Promoting Inflammation

Following the demonstration of leukocytes within neoplastic tissues by 'the father of modern cellular pathology' Rudolf Virchow in 1821–1902 [76] the concept of interconnection between inflammation, oxidative stress, cytokine production, tumour growth and metastasis has been explored multiple times [77, 78] such that inflammation is regarded as an established mechanism of carcinogenesis [78]. The major environmental factors responsible for the induction of chronic inflammation in the oral cavity are periodontal pathogens, which upregulate the host inflammatory response via activation of signalling pathways such as Mitogen-Activated Protein Kinase (MAPK)/ERK and NF-kB [79]. Thus, potential mechanisms of action of bacteria in oral cancer can be linked to the production of mediators that cause inflammation, cell proliferation, keratinocyte mutations, angiogenesis, epithelial-mesenchymal transition and subsequent invasion of transformed keratinocytes [80]. While there exists evidence of crosstalk between advanced glycation end-products and cognate receptors (RAGE) in the initiation and promotion of carcinogenesis, studies reporting induced/augmented RAGE protein expression in patients having periodontal disease strengthen the link between periodontal pathogens and OSCC [81, 82]. In response to specific molecular pattern of bacteria such as lipopolysaccharide structure, peptidoglycan type, presence of flagella and precise microbial nucleic acid sequences the host pattern recognition receptors become activated in oral epithelial cells. These include TLR (Toll-Like Receptor), a retinoic acid-inducible gene I receptor, NLRs and lectin-like receptors [83]. Nucleotide-binding oligomerization domain-like receptors or NOD-like receptors (NLRs) are intracellular sensors of microbe-associated molecular patterns (MAMPs) that enter the cell via phagocytosis or pores. Consequently, transcription factors, Nuclear Factor B kappa light chain enhancer of activated B cells (NF-kB) and mitogen-activated kinase pathways are activated, which not only activate pro-inflammatory pathways but also induce Bcl2-mediated inhibition of apoptosis, inactivation of retinoblastoma protein and stimulation of signal transducer and activator of transcription 3 (STAT3) and Nuclear Factor of Activated T cells (NFAT) [84, 85]. Eventually, this leads to uncontrolled proliferation of oral epithelial cells. Another possible mechanism in dysregulated inflammation-induced carcinogenesis is the upregulation of an enzyme named activation-induced cytidine deaminase (AID), which induces switching of the immunoglobulin gene [86]. TNF-α, produced in response to periodontal pathogens, stimulates aberrant AID expression in a non-metastatic oral cancer cell line [87]. Increased levels of TNF-α also elevate reactive oxygen species (ROS), which in turn damage DNA, as seen in all stages of tumour development from initiation to progression [88]. Thus, it may be suggested that being a predominant source of ROS, bacteria may lead to unrepaired mutations in oral epithelial cells, causing initiation and progression of oral cancer [89]. Hydrogen sulphide and nitrosamines, the direct end product of bacterial metabolism such as within *Fusobacterium nucleatum, P. gingivalis* and *P intermedia* or metabolic products of host proteins metabolised by bacteria such as *Firmicutes* and *Bacteroides* have been implicated in DNA damage through activating ROS production [90]. Similarly, another toxic metabolite, methyl mercaptan, is produced by several periodontal bacteria chiefly *Porphyromonas gingivalis* [91] which has been implicated in invasion in OSCC owing to its collagen breakdown potential. Exo U, a virulence factor in *Pseudomonas aeruginosa*, stimulates production of IL8, an inflammatory mediator by activating the NF-kB pathway [92]. There are many reports on the presence of *P. aeruginosa* in OSCC, which supports the role of these flagellated bacteria in carcinogenesis [93]. Recent reports also suggest CCL 20 mediated mechanism of *Fusobacterium nucleatum* associated oral carcinogenesis [94, 95].

9.3.2.2 Genomic Instability

Oral dysbiosis contributes to several genetic and epigenetic alterations predisposing oral epithelial cells towards carcinogenesis. Several oral bacteria such as *P. gingivalis, Rothia,* and *Streptococcus* release reactive oxygen and nitrogen species, nitrosamines, acetaldehydes and hydrogen peroxide which induce DNA damage, chromosome fragmentation and post-translational modifications in tumour suppressor genes [90, 96]. The evidence for *F. nucleatum* causing DNA damage in tongue squamous cell carcinoma cells via the Ku70/p53 pathway has been reported recently [97]. Analysis of the transcriptome profile of *F. nucleatum*-infected human immortalised oral epithelial cells demonstrated alterations in long non-coding RNA and potential hub genes [98]. Different species of bacteria have been shown to have a remarkable ability to manipulate host DNA in an epigenetic manner which the authors have named 'Pathoepigenetics' [99]. Polymicrobial infections modify the epigenetics in DNA tumour viruses to initiate carcinogenesis. Imai et al. established a link between *P. gingivalis* and the reactivation of Epstein-Barr virus in infected host cells through butyrate-induced histone deacetylation [100]. There is evidence of *P. gingivalis*-induced downregulation of enzymes leading to epigenetic regulation in HaCaT keratinocytes, thus altering the transcriptional machinery of the host cells [101].

9.3.2.3 Self-Sufficiency in Growth Signals

Aberrant growth-promoting activity may be induced by the toxic metabolites of *P. gingivalis*, which has been shown to alter human β-defensin expression in OSCC cell lines [102] leading to uncontrolled cellular proliferation. Defensin genes such as human beta-defensin 3 (HBD-3), human neutrophil peptide (HNP-1/3) and defensin-α were found to activate Epidermal Growth Factor Receptors that behave as a putative oncogene in OSCC [103, 104]. These receptors were found to be upregulated in *P. gingivalis*-infected OSCC cells [99]. *P. gingivalis* and its membrane fraction were also shown to upregulate the expression of genes involved in downstream of Toll-like Receptors (TLR), NF-kB and MAPK pathways, thus activating the pro-inflammatory immune response in gingival keratinocytes as well as malignant oral epithelial cell lines [105]. It has been shown that in vitro upregulation of *P. gingivalis*-induced matrix metalloproteinase 9 (MMP 9) upregulates p38 MAPK ERK1/2 and NKkB [106]. This supports a role for *P. gingivalis* in the continued growth of keratinocytes. Upon infection with killed *Staphylococcus aureus,* upregulation of Cyclooxygenase 2 (COX2) transcription followed by elevated expression of Cyclin was observed in human oral keratinocytes which leads to limitless cellular proliferation [107]. In vitro induction of epithelial Interleukin 6 (IL6) and TNF alpha by *F. nucleatum* and *P. gingivalis* also supports bacteria-mediated uncontrolled cell proliferation through the TLR2-NF-kB signalling pathway [64, 108]. Gram-negative bacteria secrete pro-inflammatory lipopolysaccharide endotoxin (LPS) to induce the production of IL1b, IL6 and TNF alpha by binding the TLR receptor of leucocytes [109]. *P. gingivalis* and *F. nucleatum* are shown to increase cancer cell proliferation through interleukin 6/STAT3 pathway [110]. G1 phase of cell cycle was found to be accelerated in *P. gingivalis*-infected gingival epithelial cells which promotes cell proliferation through upregulation of Cyclin D and E [111] as well as downregulation of the cell cycle kinases [112] and PTEN [113].

9.3.2.4 Evasion of Apoptosis

Loss of balance between pro-apoptotic and anti-apoptotic proteins, thus reducing the caspase activity, and altering death receptor signalling are mechanisms known to cause evasion of apoptosis. There are several shreds of evidence depicting the role of bacteria. One of the virulence factors of *E. coli,* known as a cytotoxic necrotising factor type 1 (CNF1), promotes overexpression of anti-apoptotic members of the Bcl2 gene family [114, 115]. Similarly, *Chlamydia pneumoniae,* isolated from dental plaque, resists apoptosis through downregulation of IL10 [116]. *P. gingivalis* was seen suppressing cell death in in vitro cultures of oral epithelial cells by inhibiting caspase 9 and caspase 3 [117, 118]. *P. gingivalis* has also been shown to increase the activity of STAT3 through downregulation of suppressor of cytokine signalling 3 (SOCS3), in turn inhibiting apoptosis [119]. Both *P. gingivalis* and *F. nucleatum* promote chemical-mediated oral carcinogenesis in a murine model by activating the IL6/STAT3 axis [114]. Another pathway by which *P. gingivalis* could inhibit apoptosis in the culture of gingival epithelial cells is by phosphorylating a heat shock protein HSP 27 [120].

9.3.2.5 Limitless Replicative Potential

An attribute of a cell undergoing limitless replication is the overexpression of human Telomerase Reverse Transcriptase (hTERT) leading to telomere extension. While gingival crevicular fluid and gingival biopsy tissue from healthy subjects revealed very weak or no hTERT expression, gingival tissue from patients with chronic periodontitis and with aggressive periodontitis displayed high hTERT mRNA expression at both transcription and translation levels [121]. Therefore, it may be hypothesised that bacteria implicated in periodontitis may contribute to carcinogenesis. Cells are conferred with limitless replication through telomere lengthening, as in stem cell phenotypes. Because of the existing evidence linking inflammatory cytokines such as IL1, IL6 and IL8 with cancer stem cell populations [122], bacteria producing such cytokines may be linked to carcinogenesis. In a murine model of periodontitis associated chemically induced oral carcinogenesis, *P gingivalis* and *F nucleatum promoted tumour cell proliferation* through activation of the IL6/STAT3 pathway [110].

9.3.2.6 Evading Immune Attack

Several mechanisms by which tumour cells escape immune surveillance have been reported. Programmed death ligand 1 (PD-L1)-mediated T cell exhaustion is one such mechanism found to be exploited by bacteria in carcinogenesis. *P. gingivalis* was shown to activate PD-L1/B7H1 receptors leading to anergy and apoptosis of activated T cells [123]. A molecular mechanism of *P. gingivalis*-induced upregulation of PD-L1 has been studied by Groeger et al. who concluded that internalised bacterial cell wall component peptidoglycans trigger cytosolic receptors to induce PD-L1 expression [124]. *P. gingivalis* was not only shown to inhibit phagocytosis of OSCC cells but also promoted the polarisation of macrophages into pro-tumour M2 macrophages [125].

9.3.2.7 Sustained Angiogenesis

Because cancer cells demand an elevated metabolic state to survive, angiogenesis is vital for tumour progression. Vascular endothelial growth factor (VEGF) is the key mediator in angiogenesis. Mirkeshavarz et al. [126] detected the production of IL6 and VEGF upon co-culturing cancer-associated fibroblasts and oral cancer cells. IL6 and IL8 are known to activate angiogenesis by the JAK/STAT signalling pathway [127, 128]. Activation of IL6 and IL8 by *P. gingivalis* and *F. nucleatum* raises the possibilities of bacteria-mediated angiogenesis in OSCC [129, 130].

9.3.2.8 Invasion and Metastasis

The processes of invasion followed by metastasis require cell to cell adhesion molecules (CAM) mediated epithelial-mesenchymal transition (EMT) and degradation of the extracellular matrix (ECM). To test whether periodontal pathogens could induce EMT, Abdulkareem et al. treated cultures of an OSCC cell line with heat-killed *F nucleatum*, *P. gingivalis*, and *E. coli* [131]. There was a significant activation of Snail and cytokines involved in EMT [128]. The EMT markers ZEB1 and TGF beta were found to be elevated significantly in poly-infected as well as *Fusobacterium*-infected oral cancer cell lines [132]. Production of matrix metalloproteinases MMP 1, MMP 9 and of IL8, which degrade ECM and facilitate invasion of transformed keratinocytes, is found to be increased in poly-infected oral cancer cell lines [130]. Prolonged and repetitive exposure of oral cancer cells to *P. gingivalis* results in the acquisition of EMT phenotype and enhances the production of MMP 1 and MMP 10 [133]. Furthermore *P. gingivalis* and *F. nucleatum* together triggered TLR signalling causing activation of the IL6-STAT3 axis which induced an invasive phenotype in a murine model [64]. *F. nucleatum*-infected human epithelial cells produced elevated MMP 13 following activation of several apoptosis mediating kinases such as Etk/Bmx, S6 kinase p70 and RhoA kinase p38 [134].

A detailed mechanism of Microbe-Associated Molecular Pattern (MAMP) and microbial products mediating oral carcinogenesis through swaying the eight hallmarks is described in Fig. 9.1.

9.4 Limitations of Evidence

The aforementioned studies do suffer from operational variances, chiefly the nature of controls and sample types. Adjacent normal mucosa, contralateral normal mucosa or healthy subjects were chosen as control. sample type ranges from saliva, surface swab or tissue. Moreover, technical differences, arising from different methods for lysis of bacterial cell walls, different DNA extraction techniques, choice of the hypervariable region of the 16S rRNA gene to use for amplification during PCR and the statistical pipeline used for downstream sequencing of data contribute to variability in the results obtained. Many early studies were limited resolution of the bacterial identification usually restricted to genus level. Further evidence with robust and reproducible studies is mandatory to determine whether these associations are consistent and real [135].

9.5 Conclusion

Elucidating true, functional associations between bacteria and oral cancer is challenging.

In this chapter, we have summarised the major ways through which bacteria can modulate oncogenesis. Synergistic relationships between several different bacterial species are required for the initiation and, probably different, consortia for the progression of OSCC. The oncobiomic community in oral cancer may be quite dynamic and vary between hosts. The possibility of some bacteria being protective against cancer underscores the necessity of a better understanding of the relationship between bacteria and cancer development. It is also essential not to consider bacteria separately from other microbes and to recognise individual variations in the host responses elicited.

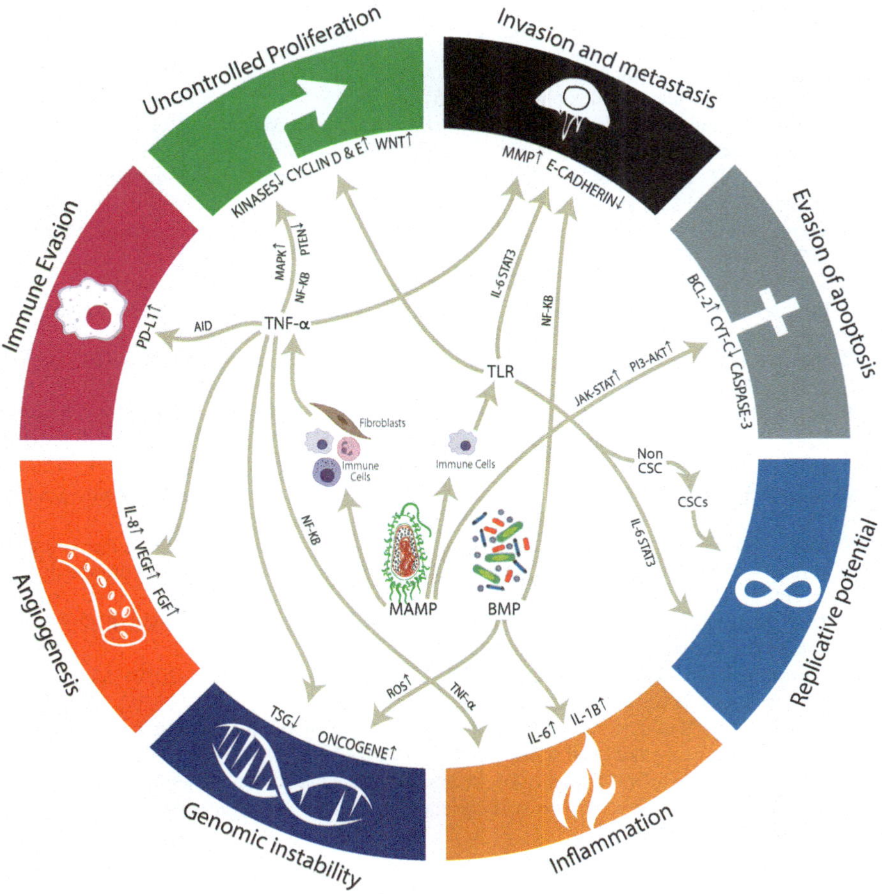

Fig. 9.1 Detailed mechanism of Microbe-Associated Molecular Pattern (MAMP) and microbial products mediated carcinogenesis through swaying the eight hallmarks. Continued cellular proliferation is chiefly achieved by MAMP-mediated downregulation of kinases and upregulation of Cyclin D, E and WNT. Invasion and metastasis are in effect due to MAMP- and BMP-mediated upregulation of matrix metalloproteinases and downregulation of E cadherin. Upregulation of Bcl2 and downregulation of cytochrome C and Caspase induced by MAMP lead to evasion of apoptosis. MAMP-induced activation of IL6-STAT3 pathways along with interaction of MAMP with cancer stem cells can lead to indefinite replicative potential of the cancer cell. Pro-inflammatory pathways such as IL6 and IL1-B are upregulated via MAMP. MAMP- and BMP-mediated upregulation of oncogenes and downregulation of TSGs can cause genomic instability. Upregulation of angiogenic factors through MAMP can cause angiogenesis. Immune evasion can take place due to MAMP-mediated PD-L1 receptor activation. *MAMP*—Microbe-Associated Molecular Pattern, *CSC*—cancer stem cell, *MMP*—matrix metalloproteinase, *WNT*—wingless-related integration site, *PD-L1*—programmed death ligand 1, *BMP*—bone morphogenetic proteins, *IL*—interleukin, *VEGF*—vascular endothelial growth factor, *FGF*—fibroblast growth factor, *TLR*—toll-like receptors, *STAT*—signal transducer and activator of transcription, *AID*—activation-induced cytidine deaminase, *ROS*—reactive oxygen species

References

1. ICD-11. https://icd.who.int/en
2. http://www.iacr.com.fr/index.php?Itemid=577
3. Ferlay J, Colombet M, Soerjomataram I, Perkin M, Pineros M, Znaor A et al (2021) Cancer statistics for the year 2020: an overview. Int J Cancer 149:778–789
4. Salehiniya H, Raei M (2021) Oral cavity and lip cancer in the world: an epidemiological review. Biomed Res Ther 7(8):3898–3905
5. Warnakulasuriya S (2009) Causes of oral cancer--an appraisal of controversies. Br Dent J 207(10):471–475
6. Johnson NW, Gupta B, Speicher DJ, Ray C, Shaikh MH, Al-Hebshi N, Gupta PC (2018) Etiology and risk factors. In: Shah J, Johnson NW (eds) Oral and oropharyngeal cancer, 2nd edn. CRC Press. https://doi.org/10.1201/9781351138543
7. Irani S (2020) New insights into oral cancer-risk factors and prevention: a review of literature. Int J Prev Med 11:202
8. Marshall B (2008) Helicobacter pylori--a Nobel pursuit? Can J Gastroenterol 22(11):895–896
9. Iebba V, Totino V, Gagliardi A, Santangelo F, Cacciotti F, Trancassini M et al (2016) Eubiosis and dysbiosis: the two sides of the microbiota. New Microbiol 39(1):1–12
10. Lederberg J, McCray A (2001) 'Ome sweet 'omics– a genealogical treasury of words. Sci Mag 15:8
11. Malla MA, Dubey A, Kumar A, Yadav S, Hashem A, Abd Allah EF (2019) Exploring the human microbiome: the potential future role of next-generation sequencing in disease diagnosis and treatment. Front Immunol 9:2868
12. Turnbaugh PJ, Ley RE, Hamady M, Fraser-Liggett CM, Knight R, Gordon JI (2007) The human microbiome project. Nature 449(7164):804–810
13. Lee-Sarwar KA, Lasky-Su J, Kelly RS, Litonjua AA, Weiss ST (2020) Metabolome-microbiome crosstalk and human disease. Meta 10(5):181
14. Gao L, Xu T, Huang G, Jiang S, Gu Y, Chen F (2018) Oral microbiomes: more and more importance in oral cavity and whole body. Protein Cell 9(5):488–500
15. Perera M, Al-hebshi NN, Speicher DJ, Perera I, Johnson NW (2016) Emerging role of bacteria in oral carcinogenesis: a review with special reference to perio-pathogenic bacteria. J Oral Microbiol 8(1):32762
16. Zaura E, Nicu EA, Krom BP, Keijser BJF (2014) Acquiring and maintaining a normal oral microbiome: current perspective. Front Cell Infect Microbiol 4:85
17. Caselli E, Fabbri C, D'Accolti M, Soffritti I, Bassi B, Mazzacane S, Franchi M (2020) Defining the oral microbiome by whole-genome sequencing and resistome analysis: the complexity of the healthy picture. BMC Microbiol 20(1):120
18. Xu X, He J, Xue J, Wang Y, Li K, Zhang K et al (2015) Oral cavity contains distinct niches with dynamic microbial communities. Environ Microbiol 17(3):699–710
19. Seidel CL, Gerlach RG, Wiedemann P, Weider M, Rodrian G, Hader M et al (2020) Defining metaniches in the oral cavity according to their microbial composition and cytokine profile. Int J Mol Sci 21(21):8218
20. Mason MR, Chambers S, Dabdoub SM, Thikkurissy S, Kumar PS (2018) Characterizing oral microbial communities across dentition states and colonization niches. Microbiome 6(1):67
21. Marsh PD, Devine DA (2011) How is the development of dental biofilms influenced by the host? J Clin Periodontol 38(Suppl 11):28–35
22. Busscher HJ, Rinastiti M, Siswomihardjo W, van der Mei HC (2010) Biofilm formation on dental restorative and implant materials. J Dent Res 89:657–665
23. Øilo M, Bakken V (2015) Biofilm and dental biomaterials. Materials 8:2887–2900
24. Wilbert SA, Mark Welch JL, Borisy GG (2020) Spatial ecology of the human tongue dorsum microbiome. Cell Rep 30(12):4003–4015.e3

25. Seerangaiyan K, van Winkelhoff AJ, Harmsen HJM, Rossen JWA, Winkel EG (2017) The tongue microbiome in healthy subjects and patients with intra-oral halitosis. J Breath Res 11(3):036010
26. Yang Y, Zheng W, Cai Q, Shrubsole MJ, Pei Z, Brucker R et al (2019) Racial differences in the oral microbiome: data from low-income populations of African ancestry and European ancestry. mSystems 4(6):e00639-19
27. Shi W, Tian J, Xu H, Zhou Q, Qin M (2018) Distinctions and associations between the microbiota of saliva and supragingival plaque of permanent and deciduous teeth. PLoS One 13(7):e0200337
28. Jose JE, Padmanabhan S, Chitharanjan AB (2013) Systemic consumption of probiotic curd and use of probiotic toothpaste to reduce Streptococcus mutans in plaque around orthodontic brackets. Am J Orthod Dentofac Orthop 144:67–72
29. De Filippis F, Vannini L, La Storia A, Laghi L, Piombino P, Stellato G et al (2014) The same microbiota and a potentially discriminant metabolome in the saliva of omnivore, ovo-lacto-vegetarian and vegan individuals. PLoS One 9:e112373
30. Wu J, Peters BA, Dominianni C, Zhang Y, Pei Z, Yang L et al (2016) Cigarette smoking and the oral microbiome in a large study of American adults. ISME J Nat Publish Group 10:2435–2446
31. Gopinath D, Wie CC, Banerjee M, Lakshmi T, Kumar P, Nallaswamy N et al (2021) Compositional profile of mucosal bacteriome of smokers and smokeless tobacco users. Clin Oral Invest. https://doi.org/10.1007/s00784-021-04137-7
32. Klaus K, Eichenauer J, Sprenger R, Ruf S (2016) Oral microbiota carriage in patients with multibracket appliance in relation to the quality of oral hygiene. Head Face Med 12:28
33. Al-Mulla A, Karlsson L, Kharsa S, Kjellberg H, Birkhed D (2010) Combination of high-fluoride toothpaste and no post-brushing water rinsing on enamel demineralization using an in-situ caries model with orthodontic bands. Acta Odontol Scand 68:323–328
34. Koopman JE, van der Kaaij NCW, Buijs MJ, Elyassi Y, van der Veen MH, Crielaard W et al (2015) The effect of fixed orthodontic appliances and fluoride mouthwash on the oral microbiome of adolescents - a randomized controlled clinical trial. PLoS One 10:e0137318
35. Jain T, Sharma P, Are AC, Vickers SM, Dudeja V (2021) New insights into the cancer-microbiome-immune axis: decrypting a decade of discoveries. Front Immunol 12:622064
36. Thorbecke G (1959) Some histological and functional aspects of lymphoid tissue in germfree animals: I. morphological studies. Ann N Y Acad Sci 78(1):237–246
37. Petersen C, Round JL (2014) Defining dysbiosis and its influence on host immunity and disease. Cell Microbiol 16(7):1024–1033
38. Meng C, Bai C, Brown TD, Hood LE, Tian Q (2018) Human gut microbiota and gastrointestinal cancer. Genomics Proteomics Bioinformatics 16(1):33–49
39. Whisner CM, Athena Aktipis C (2019) The role of the microbiome in cancer initiation and progression: how microbes and cancer cells utilize excess energy and promote one another's growth. Curr Nutr Rep 8(1):42–51
40. Gopinath D, Menon RK, Banerjee M, Su Yuxiong R, Botelho MG, Johnson NW (2019) Culture-independent studies on bacterial dysbiosis in oral and oropharyngeal squamous cell carcinoma: a systematic review. Crit Rev Oncol Hematol 139:31–40
41. Bhatt AP, Redinbo MR, Bultman SJ (2017) The role of the microbiome in cancer development and therapy. CA Cancer J Clin 67(4):326–344
42. Goodwin AC, Destefano Shields CE, Wu S, Huso DL, Wu X, Murray-Stewart TR et al (2011) Polyamine catabolism contributes to enterotoxigenic Bacteroides fragilis-induced colon tumorigenesis. Proc Natl Acad Sci U S A 108(37):15354–15359
43. Suarez G, Romero-Gallo J, Sierra JC, Piazuelo MB, Krishna US, Gomez MA et al (2017) Genetic manipulation of helicobacter pylori virulence function by host carcinogenic phenotypes. Cancer Res 77(9):2401–2412

44. Hussell T, Isaacson PG, Crabtree JE, Spencer J (1996) Helicobacter pylori-specific tumour-infiltrating T cells provide contact dependent help for the growth of malignant B cells in low-grade gastric lymphoma of mucosa-associated lymphoid tissue. J Pathol 178(2):122–127
45. Parida S, Sharma D (2021) The microbiome and cancer: creating friendly neighborhoods and removing the foes within. Cancer Res 81(4):790–800
46. Dejea CM, Wick EC, Hechenbleikner EM, White JR, Mark Welch JL, Rossetti BJ et al (2014) Microbiota organization is a distinct feature of proximal colorectal cancers. Proc Natl Acad Sci U S A 111(51):18321–18326
47. Autenrieth IB (2017) The microbiome in health and disease: a new role of microbes in molecular medicine. J Mol Med Berl Ger 95(1):1–3
48. Willis JR, Gabaldón T (2020) The human oral microbiome in health and disease: from sequences to ecosystems. Microorganisms 8(2):308
49. Nagy KN, Sonkodi I, Szöke I, Nagy E, Newman HN (1998) The microflora associated with human oral carcinomas. Oral Oncol 34(4):304–308
50. Tateda M, Shiga K, Saijo S, Sone M, Hori T, Yokoyama J et al (2000) Streptococcus anginosus in head and neck squamous cell carcinoma: implication in carcinogenesis. Int J Mol Med 6(6):699–703
51. Morita E, Narikiyo M, Yano A, Nishimura E, Igaki H, Sasaki H et al (2003) Different frequencies of Streptococcus anginosus infection in oral cancer and esophageal cancer. Cancer Sci 94(6):492–496
52. Kreth J, Merritt J, Qi F (2009) Bacterial and host interactions of oral streptococci. DNA Cell Biol 28(8):397–403
53. Hooper SJ, Crean SJ, Lewis MAO, Spratt DA, Wade WG, Wilson MJ (2006) Viable bacteria present within oral squamous cell carcinoma tissue. J Clin Microbiol 44(5):1719–1725
54. Perera M, Al-hebshi NN, Perera I, Ipe D, Ulett GC, Speicher DJ, Chen T, Johnson NW (2018) Inflammatory bacteriome and oral squamous cell carcinoma. J Dent Res 97(6):725–732
55. Yang CY, Yeh YM, Yu HY, Chin CY, Hsu CW, Liu H et al (2018) Oral microbiota community dynamics associated with oral squamous cell carcinoma staging. Front Microbiol 3(9):862
56. Gopinath D, Menon RK, Wie CC, Banerjee M, Panda S, Mandal D et al (2021) Differences in the bacteriome of swab, saliva, and tissue biopsies in oral cancer. Sci Rep 11:1181
57. Guerrero-Preston R, White JR, Godoy-Vitorino F, Rodríguez-Hilario A, Navarro K, González H et al (2017) High-resolution microbiome profiling uncovers *Fusobacterium nucleatum, Lactobacillus gasseri/johnsonii*, and *Lactobacillus vaginalis* associated to oral and oropharyngeal cancer in saliva from HPV positive and HPV negative patients treated with surgery and chemo-radiation. Oncotarget 8(67):110931–110948
58. Su Mun L, Wye Lum S, Kong Yuiin Sze G, Hock Yoong C, Ching Yung K, Kah Lok L et al (2021) Association of microbiome with oral squamous cell carcinoma: a systematic review of the metagenomic studies. Int J Environ Res Public Health 18(14):7224
59. Chang C, Geng F, Shi X, Li Y, Zhang X, Zhao X et al (2019) The prevalence rate of periodontal pathogens and its association with oral squamous cell carcinoma. Appl Microbiol Biotechnol 103(3):1393–1404
60. Katz J, Onate MD, Pauley KM, Bhattacharyya I, Cha S (2011) Presence of Porphyromonas gingivalis in gingival squamous cell carcinoma. Int J Oral Sci 3(4):209–215
61. Wen L, Mu W, Lu H, Wang X, Fang J, Jia Y et al (2020) Porphyromonas gingivalis promotes oral squamous cell carcinoma progression in an immune microenvironment. J Dent Res 99(6): 666–675
62. Geng F, Liu J, Guo Y, Li C, Wang H, Wang H et al (2017) Persistent exposure to Porphyromonas gingivalis promotes proliferative and invasion capabilities, and tumorigenic properties of human immortalized oral epithelial cells. Front Cell Infect Microbiol 7:57
63. Wu J, Zheng M, Zhang M, Pang X, Li L, Wang S et al (2018) Porphyromonas gingivalis promotes 4-nitroquinoline-1-oxide-induced oral carcinogenesis with an alteration of fatty acid metabolism. Front Microbiol 9:2081

64. Binder Gallimidi A, Fischman S, Revach B, Bulvik R, Maliutina A, Rubinstein AM et al (2015) Periodontal pathogens Porphyromonas gingivalis and Fusobacterium nucleatum promote tumor progression in an oral-specific chemical carcinogenesis model. Oncotarget 6(26): 22613–22623
65. Sawant S, Dugad J, Parikh D, Singh H (2021) Absolute quantitation of oral bacteria involved in oral cancer by real-time PCR. Med Microecol 7:100034
66. Almståhl A, Finizia C, Carlén A, Fagerberg-Mohlin B, Alstad T (2018) Mucosal microflora in head and neck cancer patients. Int J Dent Hyg 16(4):459–466
67. Rai AK, Panda M, Das AK, Rahman T, Das R, Das K et al (2021) Dysbiosis of salivary microbiome and cytokines influence oral squamous cell carcinoma through inflammation. Arch Microbiol 203(1):137–152
68. Yamashita K, Ohara M, Kojima T, Nishimura R, Ogawa T, Hino T et al (2013) Prevalence of drug-resistant opportunistic microorganisms in oral cavity after treatment for oral cancer. J Oral Sci 55(2):145–155
69. Bolz J, Dosá E, Schubert J, Eckert AW (2014) Bacterial colonization of microbial biofilms in oral squamous cell carcinoma. Clin Oral Investig 18(2):409–414
70. Zhao H, Chu M, Huang Z, Yang X, Ran S, Hu B et al (2017) Variations in oral microbiota associated with oral cancer. Sci Rep 7(1):11773
71. Al-Hebshi NN, Nasher AT, Maryoud MY, Homeida HE, Chen T, Idris AM et al (2017) Inflammatory bacteriome featuring Fusobacterium nucleatum and Pseudomonas aeruginosa identified in association with oral squamous cell carcinoma. Sci Rep 7(1):1834
72. Lee W-H, Chen H-M, Yang S-F, Liang C, Peng C-Y, Lin F-M et al (2017) Bacterial alterations in salivary microbiota and their association in oral cancer. Sci Rep 7(1):16540
73. Kageyama S, Nagao Y, Ma J, Asakawa M, Yoshida R, Takeshita T et al (2020) Compositional shift of oral microbiota following surgical resection of tongue cancer. Front Cell Infect Microbiol 10:600884
74. Pushalkar S, Mane SP, Ji X, Li Y, Evans C, Crasta OR et al (2011) Microbial diversity in saliva of oral squamous cell carcinoma. FEMS Immunol Med Microbiol 61(3):269–277
75. Gopinath D, Kunnath Menon R, Chun Wie C, Banerjee M, Panda S, Mandal D et al (2020) Salivary bacterial shifts in oral leukoplakia resemble the dysbiotic oral cancer bacteriome. J Oral Microbiol 13(1):1857998
76. David H (1988) Rudolf virchow and modern aspects of tumor pathology. Pathol - Res Pract 183(3):356–364
77. Pikarsky E, Porat RM, Stein I, Abramovitch R, Amit S, Kasem S et al (2004) NF-κB functions as a tumour promoter in inflammation-associated cancer. Nature 431(7007):461–466
78. Colotta F, Allavena P, Sica A, Garlanda C, Mantovani A (2009) Cancer-related inflammation, the seventh hallmark of cancer: links to genetic instability. Carcinogenesis 30(7):1073–1081
79. Groeger S, Jarzina F, Domann E et al (2017) Porphyromonas gingivalis activates NFκB and MAPK pathways in human oral epithelial cells. BMC Immunol 18:1
80. Niklander SE (2021) Inflammatory mediators in oral cancer: pathogenic mechanisms and diagnostic potential. Front Oral Health 22:642238. https://doi.org/10.3389/froh.2021.642238
81. Gopinath D, Kunnath Menon R, Veettil SK, George Botelho M, Johnson NW (2020) Periodontal diseases as putative risk factors for head and neck cancer: systematic review and meta-analysis. Cancers (Basel) 12(7):1893
82. Su S, Chien M, Lin C, Chen M, Yang S (2015) RAGE gene polymorphism and environmental factor in the risk of oral cancer. J Dent Res 94(3):403–411
83. Kawasaki T, Kawai T (2014) Toll-like receptor signaling pathways. Front Immunol 5:461
84. Klimesova K, Jiraskova Zakostelska Z, Tlaskalova-Hogenova H (2018) Oral bacterial and fungal microbiome impacts colorectal carcinogenesis. Front Microbiol 9:774
85. Chocolatewala N, Chaturvedi P, Desale R (2010) The role of bacteria in oral cancer. Indian J Med Paediatr Oncol Off J Indian Soc Med Paediatr Oncol 31(4):126–131
86. Okazaki I, Kotani A, Honjo T (2007) Role of AID in tumorigenesis. Adv Immunol 94:245–273

87. Nakanishi Y, Kondo S, Wakisaka N, Tsuji A, Endo K, Murono S et al (2013) Role of activation-induced cytidine deaminase in the development of oral squamous cell carcinoma. PLoS One 8(4):e62066
88. Szlosarek P, Charles KA, Balkwill FR (2006) Tumour necrosis factor-alpha as a tumour promoter. Eur J Cancer Oxf Engl 42(6):745–750
89. Choudhari SK, Chaudhary M, Gadbail AR, Sharma A, Tekade S (2014) Oxidative and antioxidative mechanisms in oral cancer and precancer: a review. Oral Oncol 50(1):10–18
90. Pang X, Tang Y, Ren X, Chen Q, Tang Y, Liang X (2019) Microbiota, epithelium, inflammation, and TGF-β signaling: an intricate interaction in oncogenesis. Front Microbiol 9:1353
91. Nakano Y, Yoshimura M, Koga T (2002) Methyl mercaptan production by periodontal bacteria. Int Dent J 52(Suppl 3):217–220. https://doi.org/10.1002/j.1875-595x.2002.tb00928.x
92. de Lima CDM, Calegari-Silva TC, Pereira RMS, de Oliveira Lima Santos SA, Lopes UG, Plotkowski M-CM et al (2012) ExoU activates NF-κB and increases IL-8/KC secretion during Pseudomonas aeruginosa infection. PLoS One 7(7):e41772
93. Al-Hebshi NN, Borgnakke WS, Johnson NW (2019) The microbiome of oral squamous cell carcinomas: a functional perspective. Curr Oral Health Rep 6(2):145–160
94. Ueda S, Goto M, Hashimoto K, Hasegawa S, Imazawa M, Takahashi M, Oh-Iwa I, Shimozato K, Nagao T, Nomoto S (2021) Salivary CCL20 level as a biomarker for oral squamous cell carcinoma. Cancer Genomics Proteomics 18(2):103–112. https://doi.org/10.21873/cgp.20245
95. Mandal DP, Mohanty N, Behera PK, Gopinath D, Panda S, Al-Kheraif AA, Divakar DD, Anil S, Panda S (2021) A plausible proposition of CCL20-related mechanism in fusobacterium nucleatum-associated oral carcinogenesis. Life 11(11):1218. https://doi.org/10.3390/life11111218
96. Meurman JH, Uittamo J (2008) Oral micro-organisms in the etiology of cancer. Acta Odontol Scand 66(6):321–326
97. Geng F, Zhang Y, Lu Z, Zhang S, Pan Y (2020) Fusobacterium nucleatum caused DNA damage and promoted cell proliferation by the Ku70/p53 pathway in oral cancer cells. DNA Cell Biol 39(1):144–151
98. Zhang S, Li C, Zhang Z, Li Y, Li Q, Geng F et al (2021) Analysis of differentially expressed genes in oral epithelial cells infected with fusobacterium nucleatum for revealing genes associated with oral cancer. J Cell Mol Med 25(2):892–904
99. Minárovits J (2009) Microbe-induced epigenetic alterations in host cells: the coming era of patho-epigenetics of microbial infections. A review. Acta Microbiol Immunol Hung 56(1):1–19
100. Imai K, Inoue H, Tamura M, Cueno ME, Inoue H, Takeichi O et al (2012) The periodontal pathogen Porphyromonas gingivalis induces the Epstein-Barr virus lytic switch transactivator ZEBRA by histone modification. Biochimie 94(3):839–846
101. de Camargo Pereira G, Guimarães GN, Planello AC, Santamaria MP, de Souza AP, Line SR et al (2013) Porphyromonas gingivalis LPS stimulation downregulates DNMT1, DNMT3a, and JMJD3 gene expression levels in human HaCaT keratinocytes. Clin Oral Investig 17(4):1279–1285
102. Hoppe T, Kraus D, Novak N, Probstmeier R, Frentzen M, Wenghoefer M et al (2016) Oral pathogens change proliferation properties of oral tumor cells by affecting gene expression of human defensins. Tumor Biol 37(10):13789–13798
103. Winter J, Pantelis A, Reich R, Martini M, Kraus D, Jepsen S et al (2011) Human beta-defensin-1, −2, and −3 exhibit opposite effects on oral squamous cell carcinoma cell proliferation. Cancer Investig 29(3):196–201
104. Wenghoefer M, Pantelis A, Dommisch H, Reich R, Martini M, Allam J-P et al (2008) Decreased gene expression of human beta-defensin-1 in the development of squamous cell carcinoma of the oral cavity. Int J Oral Maxillofac Surg 37(7):660–663

105. Groeger S, Jarzina F, Domann E, Meyle J (2017) Porphyromonas gingivalis activates NFκB and MAPK pathways in human oral epithelial cells. BMC Immunol 18(1):1. https://doi.org/10.1186/s12865-016-0185-5
106. Inaba H, Amano A, Lamont RJ, Murakami Y (2015) Involvement of protease-activated receptor 4 in over-expression of matrix metalloproteinase 9 induced by Porphyromonas gingivalis. Med Microbiol Immunol (Berl) 204(5):605–612
107. Wang Y, Liu S, Li B, Jiang Y, Zhou X, Chen J et al (2019) Staphylococcus aureus induces COX-2-dependent proliferation and malignant transformation in oral keratinocytes. J Oral Microbiol 11(1):1643205
108. Katz J, Wallet S, Cha S (2010) Periodontal disease and the oral-systemic connection: 'is it all the RAGE?'. Quintessence Int Berl Ger 41(3):229–237
109. Zhang G, Ghosh S (2000) Molecular mechanisms of NF-kappaB activation induced by bacterial lipopolysaccharide through toll-like receptors. J Endotoxin Res 6(6):453–457
110. Gallimidi AB, Fischman S, Revach B, Bulvik R, Maliutina A, Rubinstein AM et al (2015) Periodontal pathogens *Porphyromonas gingivalis* and *Fusobacterium nucleatum* promote tumor progression in an oral-specific chemical carcinogenesis model. Oncotarget 6(26): 22613–22623
111. Pan C, Xu X, Tan L, Lin L, Pan Y (2014) The effects of Porphyromonas gingivalis on the cell cycle progression of human gingival epithelial cells. Oral Dis 20(1):100–108
112. Kuboniwa M, Hasegawa Y, Mao S, Shizukuishi S, Amano A, Lamont RJ et al (2008) P. gingivalis accelerates gingival epithelial cell progression through the cell cycle. Microbes Infect 10(2):122–128
113. Al-Taweel FB, Douglas CWI, Whawell SA (2016) The periodontal pathogen Porphyromonas gingivalis preferentially interacts with oral epithelial cells in s phase of the cell cycle. Infect Immun 84(7):1966–1974
114. Fiorentini C, Matarrese P, Straface E, Falzano L, Fabbri A, Donelli G et al (1998) Toxin-induced activation of rho GTP-binding protein increases Bcl-2 expression and influences mitochondrial homeostasis. Exp Cell Res 242(1):341–350
115. Rippere-Lampe KE, Lang M, Ceri H, Olson M, Lockman HA, O'Brien AD (2001) Cytotoxic necrotizing factor type 1-positive escherichia coli causes increased inflammation and tissue damage to the prostate in a rat prostatitis model. Infect Immun 69(10):6515–6519
116. Yang J, Hooper WC, Phillips DJ, Tondella ML, Talkington DF (2003) Induction of proinflammatory cytokines in human lung epithelial cells during Chlamydia pneumoniae infection. Infect Immun 71(2):614–620
117. Yao L, Jermanus C, Barbetta B, Choi C, Verbeke P, Ojcius DM et al (2010) Porphyromonas gingivalis infection sequesters pro-apoptotic bad through Akt in primary gingival epithelial cells. Mol Oral Microbiol 25(2):89–101
118. Mao S, Park Y, Hasegawa Y, Tribble GD, James CE, Handfield M et al (2007) Intrinsic apoptotic pathways of gingival epithelial cells modulated by Porphyromonas gingivalis. Cell Microbiol 9(8):1997–2007
119. Moffatt CE, Lamont RJ (2011) Porphyromonas gingivalis induction of microrna-203 expression controls suppressor of cytokine signaling 3 in gingival epithelial cells. Infect Immun 79(7):2632–2637
120. Singh MK, Sharma B, Tiwari PK (2017) The small heat shock protein Hsp27: present understanding and future prospects. J Therm Biol 69:149–154
121. Katarkar A, Saha A, Mukherjee S, Kundu D, Bandyopadhyay P, Chaudhuri K (2015) Telomerase expression in individuals with chronic and aggressive periodontitis. J Periodontol 86(5):656–665
122. Patel S, Rawal R (2016) Role of miRNA dynamics and cytokine profile in governing CD44v6/Nanog/PTEN axis in oral cancer: modulating the master regulators. Tumour Biol 37(11): 14565–14575

123. Groeger S, Domann E, Gonzales JR, Chakraborty T, Meyle J (2011) B7-H1 and B7-DC receptors of oral squamous carcinoma cells are upregulated by Porphyromonas gingivalis. Immunobiology 216(12):1302–1310
124. Groeger S, Denter F, Lochnit G, Schmitz ML, Meyle J (2020) Porphyromonas gingivalis cell wall components induce programmed death ligand 1 (PD-L1) expression on human oral carcinoma cells by a receptor-interacting protein kinase 2 (RIP2)-dependent mechanism. Infect Immun 88(5):e00051-20
125. Liu S, Zhou X, Peng X, Li M, Ren B, Cheng G et al (2020) Porphyromonas gingivalis promotes immunoevasion of oral cancer by protecting cancer from macrophage attack. J Immunol Baltim Md 205(1):282–289
126. Mirkeshavarz M, Ganjibakhsh M, Aminishakib P, Farzaneh P, Mahdavi N, Vakhshiteh F et al (2017) Interleukin-6 secreted by oral cancer- associated fibroblast accelerated VEGF expression in tumor and stroma cells. Cell Mol Biol Noisy Gd Fr 63(10):131–136
127. Gao P, Niu N, Wei T, Tozawa H, Chen X, Zhang C et al (2017) The roles of signal transducer and activator of transcription factor 3 in tumor angiogenesis. Oncotarget 8(40):69139–69161
128. Xue C, Xie J, Zhao D, Lin S, Zhou T, Shi S et al (2017) The JAK/STAT3 signalling pathway regulated angiogenesis in an endothelial cell/adipose-derived stromal cell co-culture, 3D gel model. Cell Prolif 50(1):e12307
129. Yee M, Kim S, Sethi P, Düzgüneş N, Konopka K (2014) Porphyromonas gingivalis stimulates IL-6 and IL-8 secretion in GMSM-K, HSC-3 and H413 oral epithelial cells. Anaerobe 28:62–67
130. Proença MA, Biselli JM, Succi M, Severino FE, Berardinelli GN, Caetano A et al (2018) Relationship between fusobacterium nucleatum, inflammatory mediators and microRNAs in colorectal carcinogenesis. World J Gastroenterol 24(47):5351–5365
131. Abdulkareem AA, Shelton RM, Landini G, Cooper PR, Milward MR (2017) Periodontal pathogens promote epithelial-mesenchymal transition in oral squamous carcinoma cells in vitro. Cell Adhes Migr 12(2):127–137
132. Harrandah AM, Chukkapalli SS, Bhattacharyya I, Progulske-Fox A, Chan EKL (2021) Fusobacteria modulate oral carcinogenesis and promote cancer progression. J Oral Microbiol 13(1):1849493
133. Ha NH, Woo BH, Kim DJ, Ha ES, Choi JI, Kim SJ et al (2015) Prolonged and repetitive exposure to Porphyromonas gingivalis increases aggressiveness of oral cancer cells by promoting acquisition of cancer stem cell properties. Tumour Biol 36(12):9947–9960
134. Uitto V-J, Baillie D, Wu Q, Gendron R, Grenier D, Putnins EE et al (2005) Fusobacterium nucleatum increases collagenase 3 production and migration of epithelial cells. Infect Immun 73(2):1171–1179
135. Gopinath D, Menon RK (2021) Increasing reproducibility in oral microbiome research. Methods Mol Biol 2327:1–15

Viral-Induced Carcinopathogenesis in Oral Squamous Cell Carcinoma

10

Harpreet Kaur, Aanchal Kakkar, Sunny Kala, Muzafar A. Macha, and Deepika Mishra

Abstract

Oral cancer is multifactorial in etiology with tobacco being the most associated etiological factor. A fraction of oral cancers are associated with infectious agents, and 12% of oral cancer burden is attributed to viral etiology. Various oncogenic DNA viruses (Human Papilloma Virus (HPV), Human Herpesvirus-8 (HHV-8), Merkel cell polyomavirus (MCPV), and Hepatitis B Virus (HBV)) and oncogenic RNA viruses (Hepatitis C Virus (HCV) and human T-lymphotropic virus type 1 (HTLV-1)) have been designated as class I carcinogens by International Agency for Research on Cancer (IARC). The oncogenic viruses found to be strongly associated with oral squamous cell carcinoma (OSCC) are HPV and HSV. Other less commonly associated viruses include EBV and HCV. HPV and herpesviruses are most extensively studied "synergistic viruses" involved in OSCC. Viruses such as high-risk HPV play a role in the initiation as well progression of OSCC. Understanding the role of viruses in the pathogenesis of OSCC can help in exploring their usage as potential targets for treatment.

H. Kaur · S. Kala · D. Mishra (✉)
Division of Oral Pathology and Microbiology, Center for Dental Education and Research, All India Institute of Medical Sciences, New Delhi, Delhi, India
e-mail: hrprtarora@gmail.com; drsunnykala86@gmail.com; deepika1904@gmail.com

A. Kakkar
Department of Pathology, All India Institute of Medical Sciences, New Delhi, Delhi, India
e-mail: aanchalkakkar@gmail.com

M. A. Macha
Watson-Crick Centre for Molecular Medicine, Islamic University of Science and Technology, Pulwama, Jammu & Kashmir, India
e-mail: muzafar.aiiims@gmail.com

S. Routray (ed.), *Microbes and Oral Squamous Cell Carcinoma*,
https://doi.org/10.1007/978-981-19-0592-6_10

Keywords

Oral squamous cell carcinoma · Oncogenic virus · Carcinogens · Human papilloma virus

10.1 Introduction

Oral squamous cell carcinoma (OSCC) is defined as "a malignant epithelial neoplasm exhibiting squamous differentiation as characterized by the formation of keratin and/or the presence of intercellular bridges" [1]. It is the sixth most common cancer worldwide and more than 500,000 new cases are diagnosed each year, accounting for more than 90% of malignancies of oral cavity [2]. It is the most common malignancy in India, constituting 40–50% of all cancers [1, 3]. A high incidence of OSCC is seen in India, Hungary, Taiwan, Brazil, France, and South Africa [1]. However even after advancements in diagnostic and treatment modalities, survival rates have not improved much for decades [4]. Thus, there is a need to better comprehend the etiopathogenesis of OSCC for better interventional targeted therapy and prognostication.

10.2 Etiopathogenesis of OSCC

OSCC shows multifactorial etiopathogenesis, arising from an array of molecular alterations due to genetic susceptibility and exposure to environmental carcinogens [4]. The main environmental risk factors contributing to OSCC are tobacco, alcohol, and betel quid [5]. Alcohol in combination with tobacco intensifies the pathogenesis of OSCC and contributes to more than 75% of OSCC in developing countries. Other etiological factors are heavy metal pollutants, radiation, diet deficient in vitamins and minerals, trauma, certain bacteria, candida, poor oral hygiene, and viruses [1, 3]. Chronic exposure to these carcinogens leads to damage in genes at isolated foci or chromosomal level, leading to molecular alterations. The genetic alterations comprising activation of oncogenes and inactivation of tumor suppressor genes provide a major driving force for cells to acquire self-sufficiency in growth and evasion of growth inhibition resulting in uncontrolled cell proliferation [6, 7].

However, approximately 20% of cases lack associated risk factors. Evidence from literature indicates the role of viruses in majority of these OSCC cases [3, 8, 9]. Overall, 20% of all cancers are associated with infectious agents, and 12% of oral cancer burden is attributed to viral etiology [10, 11].

10.3 Role of Viruses in Oral Cavity Neoplasms

10.3.1 Historical Aspects

Viral theory of cancer was first proposed by Borrel in 1903. Ellerman and Bang discovered first tumor virus "the avian leukaemia virus" in 1908 followed by the isolation of Papilloma virus by Shoppe in 1933. Burkitt in 1963 identified a peculiar lymphoma caused by Epstein–Barr Virus (EBV) in children of Africa. Subsequently, Peyton Rous isolated avian sarcoma virus and found that the growth of certain bird tumors was associated with viruses. He was awarded the Nobel Prize in 1966 for his impactful discovery of tumor-inducing viruses. Almost at the same time, the US Congress launched the Special Virus Cancer Program (VCP) for research regarding the role of viruses in cancer [5, 12, 13].

10.3.2 Mechanisms of Oncogenic Viruses

Viruses may infect cells and induce a proliferative effect without causing cell death. The virus-cell complex after insertion into host DNA causes activation of oncogenes, inducing malignant transformation. In comparison to other disease-causing viruses, oncogenic viruses tend to cause persistent long-term infections [5, 12]. These oncogenic viruses contribute to carcinogenesis by direct and/or indirect mechanisms [14, 15].

Direct mechanism: In this mechanism, viruses infect the cells and express their own genes. The gene products further enhance the growth/survival of that cell. After infecting host cells, viruses are either consistently maintained as genetic elements (herpes viruses) or integrate into host DNA (retroviruses and Hepatitis B Virus (HBV)]) [14, 15].

Indirect mechanism: Viruses act as co-factors in this mechanism, without being actually present in tumor cells. They act by (1) triggering chronic inflammation-mediated oxidative damage (e.g., HBV and hepatitis C virus (HCV)) and (2) inducing immunosuppression (e.g., human immunodeficiency virus (HIV)), thereby resulting in evasion of host immunity by tumor cells [14, 15].

10.3.3 Criteria for Viral Carcinogenesis

Certain guidelines have been laid down to study the association between viruses and oral cancer. However, the most powerful tool to indicate direct association of virus and neoplasms is monoclonal analysis in the tumor. This implies that viruses being part of the initial lesion are responsible for the development of malignancy [16, 17]. The Koch's postulates described for acute transmission of acute infections cannot explain the biology of tumors caused by viruses. Thus, guidelines by Evans and Mueller were proposed to justify viral carcinogenesis.

10.3.3.1 Guidelines by Evans and Mueller [16]

Epidemiological Guidelines

- Geographic distribution of viral infection corresponds with that of tumor, adjusting for the presence of known co-factors.
- Viral markers are high in case subjects than control subjects.
- Viral markers precede tumor development, with high incidence of tumors in persons with markers than those without.
- Tumor incidence is decreased by viral infection prevention.

Virologic Guidelines

- Virus can transform cells in vitro.
- Viral genome is present in tumor cells but not in normal cells.
- Virus induces tumor in experimental animals.

10.3.3.2 Hill Criteria for Causality [17]

- Strength of association (how often is the virus associated with tumor?)
- Consistency (has association been observed repeatedly?)
- Specificity of the association (is the virus uniquely associated with tumor?)
- Temporal relationship (does viral infection precede tumorigenesis?)
- Biologic gradient (is there a dose response with viral load?)
- Biologic plausibility (is it biologically plausible that virus could cause tumor?)
- Coherence (does association makes sense with what is known about tumor?)
- Experimental evidence (is there supporting laboratory data?)

10.4 Viruses Associated with OSCC

Viruses causing human neoplasms are enumerated in Table 10.1. Oncogenic DNA viruses include EBV, HPV, Human Herpesvirus-8 (HHV-8), Merkel cell polyomavirus (MCPV), and HBV. HCV and human T-lymphotropic virus type 1 (HTLV-1)

Table 10.1 Overview of early (E) and late (L) gene products of HPV [18]

Gene product	Description
E1	Controls viral replication and gene transcription—Helicase function
E2	Controls viral replication and gene transcription—Viral transcription factor
E4	Viral assembly—Interact with cytoskeleton
E5	Growth induction, downregulate HLA-1 molecule
E6	Degradation of p53, antiapoptotic, activation of telomerase and cell immortalization, genomic instability
E7	Transactivates E2F-dependent promoters, genomic instability, cell immortalization
L1	Major capsid protein
L2	Minor capsid protein, nuclear transport of viral DNA, and encapsidation of viral genomes

are oncogenic RNA viruses. These oncogenic viruses are designated as class I carcinogens by International Agency for Research on Cancer (IARC). The oncogenic viruses associated with OSCC are divided into two categories: (1) strongly associated with OSCC: Human Papilloma Virus (HPV) and Herpes Simplex Virus (HSV); and (2) less strongly associated with OSCC: EBV and Hepatitis C Virus (HCV). HPV and herpes viruses are most extensively studied "synergistic viruses" involved in OSCC [5, 16, 18].

10.4.1 Human Papilloma Virus and OSCC

Structure: HPV family belongs to the family of Papovaviridae and consists of more than 200 genotypes, of which more than 30 have been known to affect oral mucosa. They are DNA viruses which target basal cells of ectoderm of skin and epithelium of oral mucosa. They are non-enveloped icosahedral viruses, 8 kbp in length with double-stranded circular DNA containing almost 8000 base pairs and 8 open reading frames (ORF). The HPV genome comprises three regions consisting of first noncoding region called upstream regulatory protein (URR) and long control region (LCR) which regulates transcription of E6 and E7 viral genes. The second region is early (E) region encoding E1-E7 proteins, and the third region is late (L) region which encodes for late proteins L1-L2. Early proteins E1-E5 are nonstructural proteins responsible for the replication and transcription of genome. E6-E7 proteins are involved in tumoral transformation of host cells. The late proteins L1-L2 encode for structural proteins of viral capsid [1, 18] (Table 10.1).

Types: HPV are divided into low-risk and high-risk types. Low-risk (LR) types include types 6, 11, 13, 32, 40, 42, 43, 44, 54, 55, 61, 62, 64, 67, 69, 70, 71, 72, and 81 which are responsible for benign lesions such as warts (Papilloma, verruca vulgaris, condyloma accumunatum, Heck's disease, etc.). High-risk (HR) types include 6, 18, 31, 33, 35, 39, 45, 51, 52, 56, 58, 59, 66, 68, and 73 which are associated with malignant neoplasms. Syrjanen in 1983 first commended HPV as a causative agent of head and neck cancer which was evidenced by (1) established role of HPV in cervical cancer, (2) similarity in clinical features of genital and oral injuries, (3) similarities in epithelia of genitals and oral cavity, (4) affinity of HPV to epithelium and its malignant potential and (5) detection of HPV in OSCC samples [1, 18, 19].

Prevalence and Site Characteristics: There are discrepancy in frequency distribution of HPV in OSCC in different studies. The variability can not only be attributed to different techniques used for HPV detection (sampling method, storage conditions, and divergent methodology of PCR or ISH for detection) but also by the lesion site (oropharynx, tongue, or floor of mouth). HPV is predominantly seen in association with SCC of tonsil and oropharynx than other head and neck sites. Syrjanen et al. performed a systematic review and found important association with HPV 16 and OSCC with 33.7% prevalence in OSCC [20]. One study by St Guily et al. showed 10.5% prevalence of HPV in OSCC and 46.5% in oropharyngeal SCC, which showed female predominance. HPV16 was most prevalent and was found in around 96% of HPV positive cases. The sites most frequently involved were tonsil (58.9%) and base of tongue (13.7%) in oropharynx; and floor of mouth (41.1%) and oral part of tongue (38.3%) in oral cavity SCC [21].

10.4.1.1 Pathogenesis

The pathogenesis of HPV-induced benign, premalignant, and malignant lesions is depicted in Figs. 10.1–10.3. The viral protein E6 promotes degradation of p53 and E7 causes inactivation of retinoblastoma (Rb) protein causing dysregulation of cell cycle leading to malignant transformation [19–22].

10.4.1.2 Demonstration of HPV

HPV causes proliferation of epithelial cells with increased epithelial thickening, prominent stratum granulosum, acanthosis, and hyperkeratosis sometimes. The squamous epithelial cells exhibit perinuclear clearing and cytoplasm with increased density, which are called koilocytes. These koilocytes represent cytopathic effect of HPV and act as clue to diagnosis of HPV infection. Increased prevalence of HPV is seen in OSCC in different tumor grades. The methods used for detection are p16 immunohistochemistry, in situ hybridization, and polymerase chain reaction [23, 24]. Recent studies have shown that HPV-positive OSCC have intact p16 and

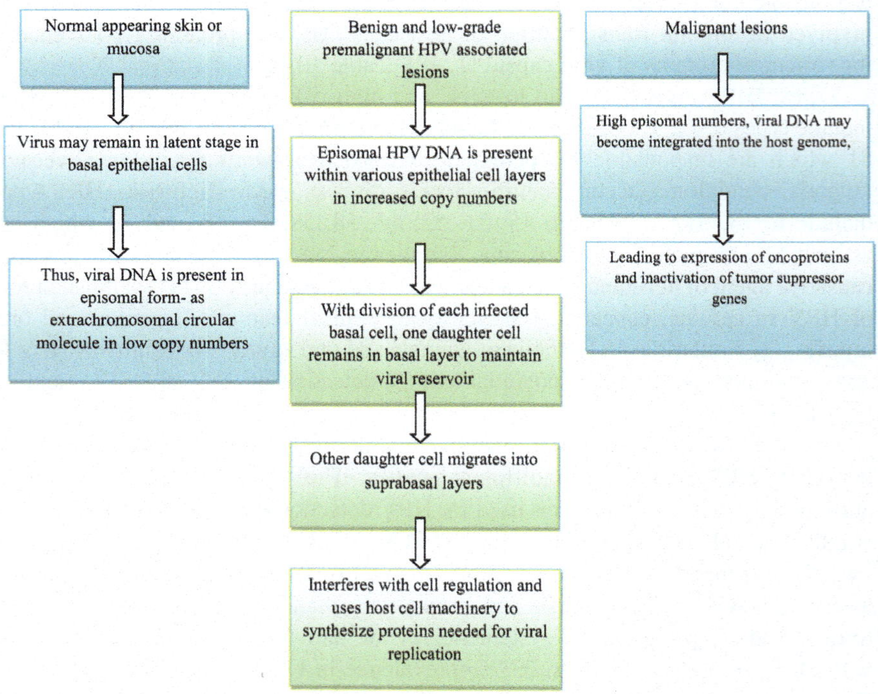

Fig. 10.1 HPV in normal mucosa, benign and premalignant, and malignant lesions

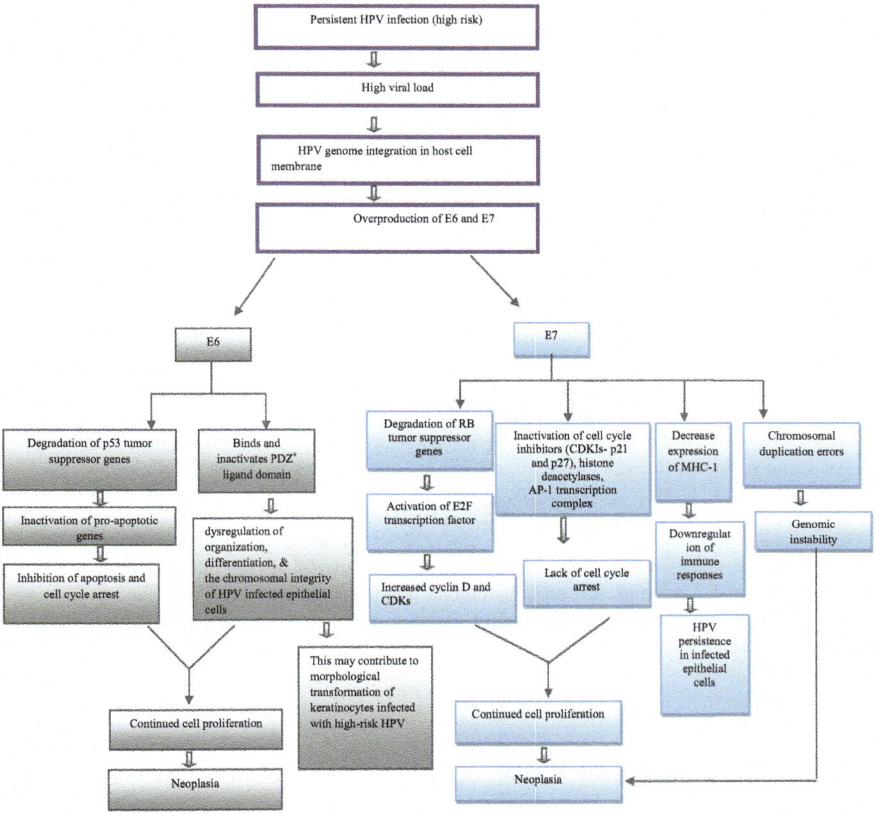

Fig. 10.2 Pathogenesis of HPV-induced carcinogenesis. *CDK* cyclin-dependent kinases, *CDKI* cyclin-dependent kinase inhibitors, *MHC* major histocompatibility complex. *Asterisk*, *PDZ* is an initialism combining the first letters of the first three proteins discovered to share the *domain*—postsynaptic density protein (PSD95), Drosophila disc large tumor suppressor (Dlg1), and zonula occludens-1 protein (zo-1). The PDZ domain is located at areas of cell-to-cell contact, such as tight junctions of epithelial cells, and is associated with signal transduction pathways

wild-type p53 gene and thus p16 protein detection by immunohistochemistry could be used for HPV detection [18].

10.4.1.3 Prognosis of HPV-Associated OSCC

HPV-associated OSCC is more sensitive to radiation and shows better prognosis as compared to OSCC with negative HPV status. Analysis of molecular mechanisms revealed that HPV-associated OSCC has a lesser degree of chromosomal aberrations and aneuploidy than HPV negative OSCC [19, 24].

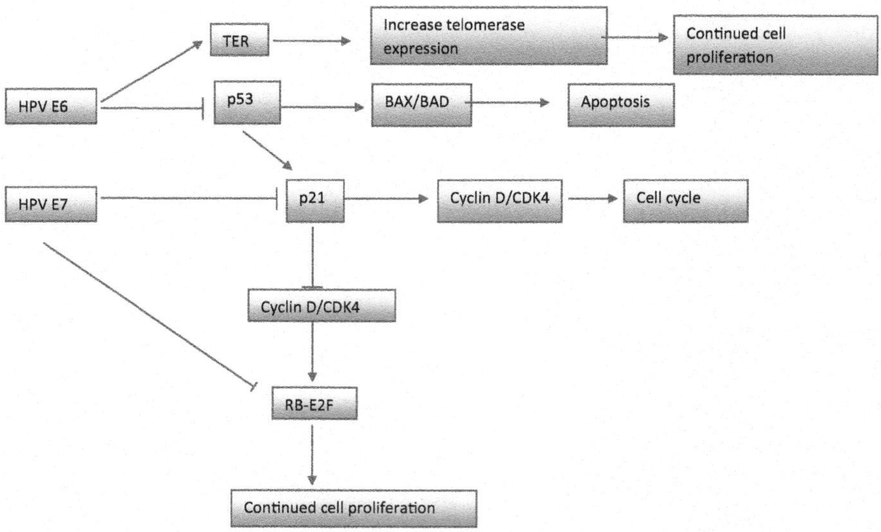

Fig. 10.3 Pathogenesis of HPV-induced carcinogenesis. HPV causes malignant transformation through E6 and E7 proteins. E6 protein inhibits apoptosis by inhibiting p53 (tumor suppressor gene). It also induces increased telomerase expression causing continued cell proliferation. E7 protein causes inhibition of p21 and retinoblastoma (RB) gene leading to uninterrupted cell proliferation

10.4.2 Herpes Simplex Virus

Herpes simplex virus (HSV) is an enveloped spherical virus with double-stranded DNA. Two forms of virus have been recognized, with HSV-1 associated with oral and ocular lesions and HSV-2 mainly causing genital lesions. The genome of HSV-1 is tightly packed in a phage-like liquid crystalline manner enveloped by icosahedral capsid which is 100 nm. However, overall structure and behavior of these two viruses are more or less similar [1]. Both the viruses cause primary infection followed by latent infection that may be succeeded by secondary recurrent infections.

10.4.3 HSV in Oral Carcinogenesis

The first animal model of HSV acting as a co-carcinogen was demonstrated by Southam et al. The malignant transformation mechanism of HSV is not clear; however, even a small amount of viral DNA is not retained by cells, thereby justifying the hit and run mechanism of transformation [10]. Some authors suggest that HSVs transform cells by acting as a mutagen. These mutations cause microsatellite instability leading to genomic alterations. Other mechanisms include induction of cell proteins [10], host cell shut off process [25], by stimulation of other viruses [26], and targeting chromosomes [27].

Induction of cell proteins: Heat shock proteins are induced by HSV infection which might be responsible for cell transformation.

Shutting off host cell process: The infected cell halts synthesis of cellular proteins and rapid degradation of RNA might be associated with HSV-mediated cellular transformation.

Induction of other viruses: As in cervical carcinogenesis, HSV and HPV act as co-carcinogens with HSV acting as initiator and HPV as promoter. However, in oral lesions very few carcinoma tissues appeared to have both HSV 1 and HPV 16. Nevertheless, synergistic effect between these two viruses and OSCC cannot be disregarded.

Chromosome as targets: Damage to specific chromosomal sites leads to chromosomal rearrangement which may be a mechanism for HSV-mediated transformation.

10.4.4 Epstein–Barr Virus in OSCC

Epstein–Barr Virus (EBV) was discovered by Michael Anthony Epstein in 1964. It becomes latent in B-lymphocytes and thus is called as lymphocryptovirus. EBV belongs to gamma subfamily of Herpes viruses. It is involved in a variety of neoplasms, namely lymphomas, nasopharyngeal carcinoma, lymphoepithelial-like carcinoma, and undifferentiated carcinoma [11]. Proliferative verrucous leukoplakia, a form of leukoplakia which has a very high rate of malignant transformation to OSCC, has a strong association with EBV [1, 28].

EBV in oral cancer: The role of EBV in OSCC has been controversial. A study from Taiwan revealed that majority (82.5%) of specimens were EBV positive. Zheng et al. and Gonzalez-Moles et al. had shown increased EBV positivity with decreased degree of differentiation and higher grade of OSCC [18, 29]. Delavarian et al. reported association between EBV and OSCC and concluded that in addition to tobacco and alcohol, EBV also plays an important role in the pathogenesis of OSCC, especially in patients younger than 40 years [30]. Horuchi et al. also demonstrated EBV in epithelial cells of OSCC, carcinoma in situ, and leukoplakia [31].

10.4.5 Hepatitis C Virus and OSCC

HCV is a single-stranded, enveloped RNA virus isolated from chimpanzee in 1989. It is most commonly associated with chronic liver disease and also causes extrahepatic manifestations which involve oral cavity as well [32]. Many studies in literature support that HCV is strongly associated with the etiology of oral lichen planus (OLP); however pathogenic association of HCV with OLP is controversial [33]. Nagao et al. had found anti-HCV antibodies in the sera of OSCC patients. As squamous cells in oral cavity are exposed to HCV from saliva and serum in HCV-positive patients, that might possibly explain the development of OLP and OSCC [34]. Su et al. had found causal association of HCV and OSCC [35]. However some authors suggest that high HCV antibody titer in patients with OSCC was linked to older age and may not have any association with the etiology of oral cancer [36].

10.5 Conclusion

Oral cancer is multifactorial in etiology with tobacco smoking being the most commonly associated etiological factor. The role of viruses in the etiopathogenesis of OSCC has been studied in literature with considerable interest especially in nonsmokers. Viruses such as high-risk HPV play a role in the initiation as well as progression of OSCC, which histopathologically shows non-keratinizing epithelium with characteristic koilocytes. Other viruses associated with OSCC are HSV, EBV, and HCV; however, their specific role in pathogenesis has to be elucidated in detail. These viruses can act as potential targets for treatment and widen our horizon of understanding the pathogenesis of OSCC.

References

1. Neville B, Damm DD, Allen C, Chi A (2015) Oral and maxillofacial pathology, 4th edn. Elsevier, St. Louis. https://www.elsevier.com/books/oral-and-maxillofacial-pathology/neville/978-1-4557-7052-6. Accessed 18 Dec 2020
2. Irimie AI, Braicu C, Cojocneanu-Petric R, Berindan-Neagoe I, Campian RS (2015) Novel technologies for oral squamous carcinoma biomarkers in diagnostics and prognostics. Acta Odontol Scand 73(3):161–168
3. Chaturvedi P, Chocolatewala N (2009) Role of human papilloma virus in the oral carcinogenesis: an Indian perspective. J Cancer Res Ther 5(2):71
4. Warnakulasuriya S (2014) Prognostic and predictive markers for oral squamous cell carcinoma: the importance of clinical, pathological and molecular markers. Saudi J Med Med Sci 2(1):12
5. Patil SP, Sodhi SJ, Tambe SD, Gada V, Vikhe D (2016) Viruses in oral squamous cell carcinoma: a review. Pravara Med Rev 8(1):4–7
6. Choi S, Myers JN (2008) Molecular pathogenesis of oral squamous cell carcinoma: implications for therapy. J Dent Res 87(1):14–32
7. Kumar V, Abbas AK, Fausto N, Aster JC (2014) Robbins and Cotran pathologic basis of disease. Elsevier, Philadelphia, p 1413p
8. Venuti A, Badaracco G, Rizzo C, Mafera B, Rahimi S, Vigili M (2004) Presence of HPV in head and neck tumours: high prevalence in tonsillar localization. J Exp Clin Cancer Res 23(4): 561–566
9. Li W, Thompson CH, Cossart YE, O'Brien CJ, McNeil EB, Scolyer RA et al (2004) The expression of key cell cycle markers and presence of human papillomavirus in squamous cell carcinoma of the tonsil. Head Neck 26(1):1–9
10. Metgud R, Astekar M, Verma M, Sharma A (2012) Role of viruses in oral squamous cell carcinoma. Oncol Rev 6(2)

11. Mui UN, Haley CT, Tyring SK (2017) Viral oncology: molecular biology and pathogenesis. J Clin Med 29(6):111. https://www.ncbi.nlm.nih.gov/pmc/articles/PMC5742800/ Accessed 20 Dec 2020
12. Gayathri C, Sivaramakrishnan M, Suganya R, Santhadevy A, Vezhavendhan N (2019) Pathogenesis of oral squamous cell carcinoma—an update. Int Dent J Stud Res 7(4):84–86
13. Javier RT, Butel JS (2008) The history of tumor virology. Cancer Res 68(19):7693–7706
14. Morales-Sánchez A, Fuentes-Panana EM (2014) Human viruses and cancer. Viruses 6:4047–4079
15. Mui UN, Haley CT, Tyring SK (2017) Viral oncology: molecular biology and pathogenesis. J Clin Med 6(12):111. https://www.ncbi.nlm.nih.gov/pmc/articles/PMC5742800/ Accessed 20 Dec 2020
16. Evans AS, Mueller NE (1990) Viruses and cancer. Causal associations. Ann Epidemiol 1(1): 71–92
17. Hill AB (1965) The environment and disease: association or causation? Proc R Soc Med 58(5): 295–300
18. Gupta K, Metgud R (2013) Evidences suggesting involvement of viruses in oral squamous cell carcinoma. Patholog Res Int 2013:642496. https://doi.org/10.1155/2013/642496
19. Ajila V, Shetty H, Babu S, Shetty V, Hegde S (2015) Human papilloma virus associated squamous cell carcinoma of the head and neck. J Sex Transm Dis 2015:791024. https://doi.org/10.1155/2015/791024
20. Syrjänen S, Lodi G, von Bültzingslöwen I, Aliko A, Arduino P, Campisi G et al (2011) Human papillomaviruses in oral carcinoma and oral potentially malignant disorders: a systematic review. Oral Dis 17(Suppl 1):58–72
21. St Guily JL, Jacquard A-C, Prétet J-L, Haesebaert J, Beby-Defaux A, Clavel C et al (2011) Human papillomavirus genotype distribution in oropharynx and oral cavity cancer in France—The EDiTH VI study. J Clin Virol 51(2):100–104
22. Wołącewicz M, Becht R, Grywalska E, Niedźwiedzka-Rystwej P (2020) Herpesviruses in head and neck cancers. Viruses 12(2):172. https://doi.org/10.3390/v12020172
23. Raj D (2019) Oncogenic viruses in oral neoplasia. Int J Dent Oral Heal 5(5):68–74. https://www.scribd.com/document/475768844/Oncogenic-Viruses-In-Oral-Neoplasia-pdf. Accessed 17 Jan 2021
24. Sathish N, Wang X, Yuan Y (2014) Human papillomavirus (HPV)-associated oral cancers and treatment strategies. J Dent Res 93(7 Suppl):29S
25. Steele C, Shillitoe EJ (1991) Viruses and oral cancer. Crit Rev Oral Biol Med 2(2):153–175
26. Boyd AL, Enquist L, Vande Woude GF, Hampar B (1980) Activation of mouse retrovirus by herpes simplex virus type 1 cloned DNA fragments. Virology 103(1):228–231
27. Stich HF, Hsu TC, Rapp F (1964) Viruses and mammalian chromosomes. I. Localization of chromosome aberrations after infection with herpes simplex virus. Virology 22:439–445
28. She Y, Nong X, Zhang M, Wang M (2017) Epstein-Barr virus infection and oral squamous cell carcinoma risk: a meta-analysis. PLoS One 12(10):e0186860
29. Zheng Y, Xia P, Zheng H-C, Takahashi H, Masuda S, Takano Y (2010) The screening of viral risk factors in tongue and pharyngolaryngeal squamous carcinoma. Anticancer Res 30(4): 1233–1238
30. Delavarian Z, Pakfetrat A, Falaki F, Pazouki M, Pazouki N (2010) The role of viruses in oral squamous cell carcinoma in young patients in Khorasan (Northeast of Iran). J Appl Sci 10(11): 981–985. https://scialert.net/abstract/?doi=jas.2010.981.985. Accessed 18 Jan 2021
31. Horiuchi K, Mishima K, Ichijima K, Sugimura M, Ishida T, Kirita T (1995) Epstein-Barr virus in the proliferative diseases of squamous epithelium in the oral cavity. Oral Surg Oral Med Oral Pathol Oral Radiol Endod 79(1):57–63. https://doi.org/10.1016/s1079-2104(05)80075-7. PMID: 7614163
32. Johnson RJ, Gretch DR, Yamabe H, Hart J, Bacchi CE, Hartwell P et al (1993) Membranoproliferative glomerulonephritis associated with hepatitis C virus infection. N Engl J Med 328(7):465–470

33. Lavanya N, Jayanthi P, Rao UK, Ranganathan K (2011) Oral lichen planus: an update on pathogenesis and treatment. J Oral Maxillofac Pathol 15(2):127
34. Nagao Y, Sata M, Itoh K, Tanikawa K, Kameyama T (1996) Quantitative analysis of HCV RNA and genotype in patients with chronic hepatitis C accompanied by oral lichen planus. Eur J Clin Invest 26(6):495–498
35. Su F-H, Chang S-N, Chen P-C, Sung F-C, Huang S-F, Chiou H-Y et al (2012) Positive association between hepatitis C infection and oral cavity cancer: a nationwide population-based cohort study in Taiwan. PLoS One 7(10):e48109
36. Takata Y, Takahashi T, Fukuda J (2002) Prevalence of hepatitis virus infection in association with oral diseases requiring surgery. Oral Dis 8(2):95–99

Fungi: Culprits or Bystanders in Oral Squamous Cell Carcinoma

11

Jai Ranjan, Vinaykumar Hallur, and Shivaprakash M. Rudramurthy

Abstract

Fungi are mainly saprophytic organisms but can be implicated as a causative factor for certain diseases. Of late, various fungi have also been found to have a causal relationship in various carcinomas, with aflatoxin being predominantly associated with hepatocellular carcinoma. Similarly, studies also have shown that oral potentially malignant lesions (OPMD) and oral squamous cell carcinomas (OSCC) can be caused due to fungal infections of the oral cavity. It is mainly *Candida* spp. that has been found to induce oral carcinopathogenesis. This is proposed to be multifactorial with production of hyphal forms and candidalysin by *Candida* spp.

This chapter discusses the possible association of oral cancer with fungal infections of the oral cavity. The various factors influencing fungal infections in oral cancers and the causal relationship between fungal infections and oral carcinopathogenesis are also elaborated.

Keywords

Fungi · OPMD · OSCC · *Candida*

J. Ranjan · V. Hallur
Department of Microbiology, All India Institute of Medical Sciences (AIIMS), Bhubaneswar, Bhubaneswar, Odisha, India
e-mail: jairanjan5@gmail.com; vinay118@gmail.com

S. M. Rudramurthy (✉)
Division of Mycology, Department of Medical Microbiology, Postgraduate Institute of Medical Education and Research, Chandigarh, Chandigarh, India
e-mail: mrshivprakash@yahoo.com

S. Routray (ed.), *Microbes and Oral Squamous Cell Carcinoma*,
https://doi.org/10.1007/978-981-19-0592-6_11

11.1 Introduction

Fungi are eukaryotic organisms that can be unicellular or multicellular. They are mostly found as saprobes in indoor or outdoor environment, but many species can cause infections in a susceptible host [1]. Certain fungi which produce toxins, particularly those producing aflatoxin B1, are implicated in causing cancerous lesions in the liver [2]. Aflatoxin B1 results in 249ser mutation, which inhibits the action of p53 resulting in carcinogenic effects [3]. Oral candidiasis caused by *Candida* spp. is also associated with malignant lesions in the oral cavity [4]. This chapter will discuss the possible association of oral cancer with fungal infections of the oral cavity. The various factors influencing fungal infections in oral cancers and the causal relationship will also be dealt with in this section.

Oral cancer is one of the most common malignancies reported worldwide. Its incidence is about 400,000 cases, with the majority of cases from South-East Asian countries like India [5]. Squamous cell carcinomas are the predominant type, with its cases amounting to about 90% of all the described types of oral cancer [6]. Surgery, radiotherapy, and chemotherapy are preferred treatment modalities for oral carcinoma.

Tobacco chewing, betel nut and alcohol consumption are the risk factors for oral cancer [7]. Epithelial dysplasia, resulting from chronic inflammation, is usually the first sign of oral cancerous lesions. These dysplastic cells can slowly transform into malignant cells and migrate to distant sites leading to metastasis [8]. Dysplasia is thought to be a result of long-term inflammation due to various infectious agents, which result in the production of various cytokines such as interleukins (IL)-1α, IL-1β, tumor necrosis factor (TNF)-α, etc. [9].

11.2 *Candida* Species and Oral Cavity

The oral carriage of *Candida* spp. ranges between 2 and 69% among the healthy population [10, 11]. It is reported that oral colonization of *Candida* increases to about 75% after chemotherapy and to about 93% after radiotherapy [12]. The presence of *Candida* in healthy individuals presents a challenge to epidemiological studies on the association of *Candida* with oral cancers; nevertheless, a correlation between *Candida*, oral potentially malignant conditions, and oral cancers was found in some studies.

11.2.1 *Candida* spp. and Potentially Malignant Oral Disorders

WHO in 2007 coined the term "oral potentially malignant disorders" (OPMDs) for precancerous lesions and conditions of the oral cavity [13]. Of the various OPMDs, studies have mainly focused on the association of *Candida* spp. with oral leucoplakia (OL) and its progression to oral cancer. Yeasts reported from these lesions are *C. albicans*, *C. glabrata*, *C. tropicalis*, *Candida pintolopesii*, and *S. cerevisiae*

[14]. Shukla et al. conducted a systematic review on *Candida* infection and OL and identified 16 studies using rigorous criteria. After analyzing the data from those studies, they found that the incidence of *Candida* infection in OL ranged from 6.8 to 100% [15]. Only three studies reported OL with Candida's malignant transformation rates, which were 2.5, 6.5, and 28.7% [16–18]. The remaining 13 studies provided data only on the association of *Candida* with epithelial dysplasia and cellular atypia, which may progress to cancer. Between 20 and 100% of such patients had dysplasia which ranged from mild to carcinoma-in-situ. One of the studies analyzed had reported higher grades of dysplasia (50%) in patients with *Candida* compared to those without *Candida* infections (29.2%) [19]. However, they acknowledge several limitations in their analysis. The most important of the limitations was the lack of uniformity of definitions, affecting the consistency of analysis and non-recruitment of studies on chronic hyperplastic candidiasis. In older studies, oral leukoplakia was considered a synonym of chronic hyperplastic candidiasis (CHC). However, these are now recognized as two different entities. CHC presents as adherent white patches on the oral mucosa, and on histological examination, parakeratosis with hyphal forms invading the epithelium is found [20]. Cawson, in his study, found that six of ten patients with an initial diagnosis of CHC developed oral squamous cell carcinoma (OSCC) [21]. Similarly, Eyre and Nally found that two of three patients with CHC transformed to OSCC [22]. Further, some patients responded to antifungal therapy with the resolution of symptoms, strengthening the role of *Candida* in carcinogenesis [23].

11.2.2 *Candida* spp. and Oral Cancer

Candida is associated with mucosal cancer since the 1980s. Krogh et al., in their study, showed that certain *Candida albicans* strains (biotypes 051, 147, 151, 153, 157 and 353) have a greater ability to form N-Nitrosobenzylmethylamine (NBMA) from its precursors. More such strains of *C. albicans* were isolated from more advanced precancerous lesions, suggesting a causal relationship between them [24]. McCullough et al. showed a significant association between oral epithelial dysplasia with carriage of yeasts in the oral cavity [25]. It is mainly *C. albicans* that has been found to have a significant association with OSCC. A study by Alnuaimi et al. showed that *C. albicans* genotype A is mainly found in oral cancer patients [26]. In a recent meta-analysis by Patil et al., a significant association was found between OSCC and *Candida* or *C. albicans* infection. However, considering the fact that the analysis included a limited number of studies and that these studies had lower Newcastle-Ottawa scores, more studies are needed to confirm this association [27]. In a recent study by Sankari et al., the odds ratio of oral candida carriage with oral squamous cell carcinomas (OSCC) was 11.87 [28]. However, despite the above evidence, the causal association of *Candida* is yet to be established, and larger longitudinal studies are required for conclusively proving its role in oral cancer.

11.2.3 Oral Microbiome

The oral microbiome is mainly composed of bacteria such as *Streptococcus* spp., facultative anaerobes like *Actinomyces* and obligate anaerobes, i.e., *Fusobacterium* and *Veillonella* and fungi, the majority of which belong to *Candida* spp. [29]. Other fungi which are members of the oral mycobiome belong to *Cladosporium,* Saccharomycetales, *Aureobasidium, Aspergillus, Malassezia, Fusarium, Cryptococcus,* etc. [30]. In recent studies, the oral microbiome has also been incriminated with the causation of oral cancer [31]. Shay et al. conducted a case-control study on the oral wash samples of patients with head and neck cancers (cases) and healthy individuals (controls). Interestingly, approximately 59% of their cases had oral cancer. They reported that the bacteria *R. mucilaginosa* and yeast *C. albicans* were abundant in patients with head and neck cancers [32]. A similar study was done by Perera et al., where patients with OSCC were compared with patients with fibroepithelial polyps. It found an association of altered mycobiome with over-representation of *Candida, Hannaella, Gibberella* and an elevated level of *C. albicans* with oral cancer. Other salient findings of their study were the significant association of *C. etchellsii* with OSCC and the presence of fungal taxa capable of producing anticancer compounds exclusively in non-cancerous tissues [30].

11.2.4 *C. albicans* and Oral Cancer Pathogenic Mechanisms

Candida is equipped with many virulence factors that help in colonization of the normal oral cavity and OPMD lesions, which may or may not progress to oral cancer. Promotion of oral cancer by *Candida* is by the interaction of multiple factors such as virulence factors of *Candida*, host and environmental factors. After colonization, *Candida* degrades the keratinocytes by either releasing secretory proteases enzymes that degrade and digest the oral tissue [33] or get endocytosed in keratinocytes by the E-cadherin pathway [34]. The Als3 protein present on the surface of *Candida* interacts with E-cadherin on the host cells resulting in fungal invasion into the host cells by endocytosis [34]. This endocytosis can result in damage to the epithelial cell layer and hence can lead to carcinogenesis [35]. Figure 11.1. shows the various factors implicated in carcinogenesis due to *C. albicans*.

11.2.5 *Candida* Promotes Oral Cancer

Dwivedi et al. evaluated the effect of a synthetic carcinogen, i.e., 4-Nitroquinoline 1-oxide (4NQO), along with the presence of *C. albicans* in immunocompetent mice. These two factors in combination were shown to result in dysplastic lesions, which could eventually turn malignant. None of the mice in the control group, i.e., mice that were not administered 4NQO, developed dysplastic lesions [36]. This signifies that a combination of *C. albicans* along with synthetic carcinogen (4NQO) is necessary for the development of dysplasia. The interaction between 4NQO and

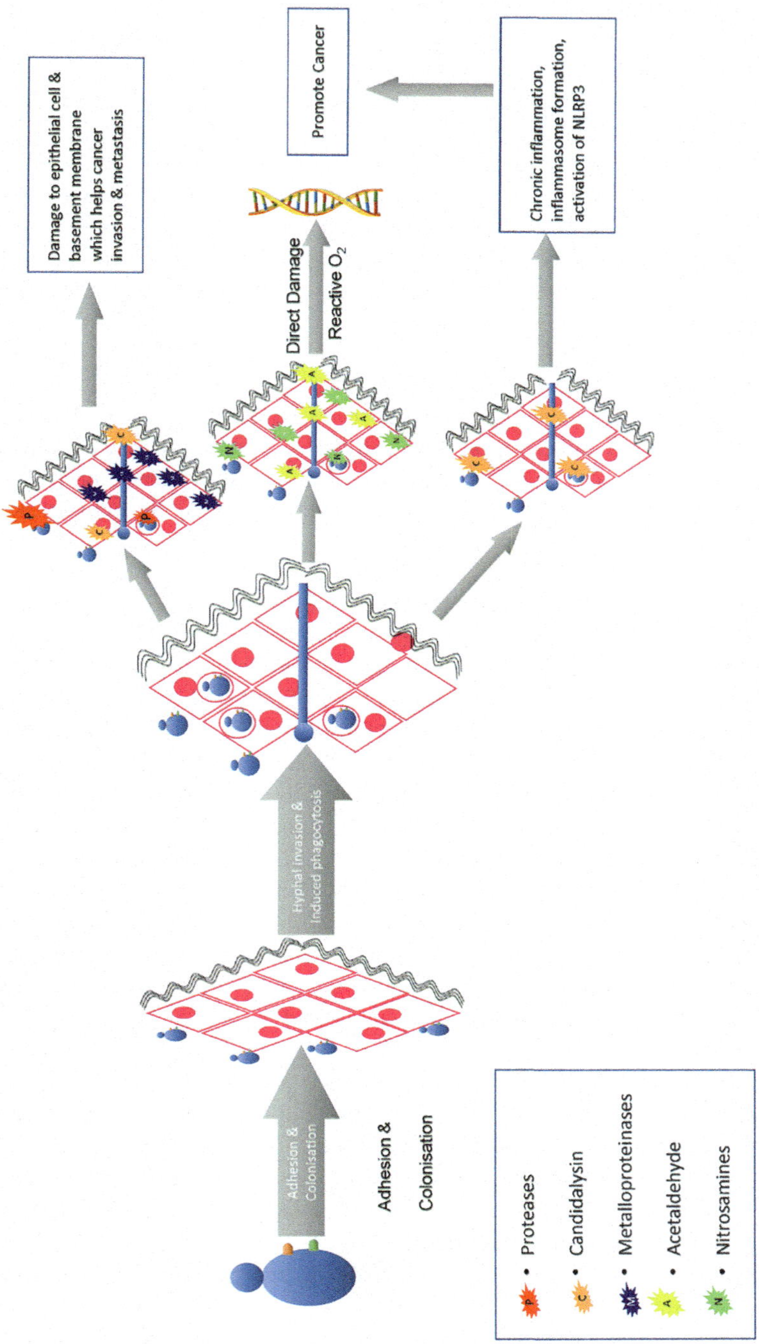

Fig. 11.1 Factors implicated in carcinogenesis due to *Candida albicans*

C. albicans results in increased expression of p16 and Ki-67. These, in turn, might lead to the proliferation of genetically altered cells [37].

11.2.5.1 Acetaldehyde

Tobacco and alcohol consumption are some of the most critical risk factors for oral squamous cell carcinoma (OSCC) [38]. Alcohol or ethanol is metabolized to acetaldehyde which is a toxic compound [39]. The acetaldehyde concentration in saliva was shown to increase in alcoholics who were also using tobacco products [40]. *C. albicans* also produces an alcohol dehydrogenase (ADH) which is encoded by the CaADH1 gene. The CaAdh1p enzyme produced by *C. albicans* can increase acetaldehyde in the oral cavity [41]. This can, in turn, act as an agent in the causation of oral cancers.

11.2.5.2 Hyphal Forms and Candidalysin

The virulence of *Candida* is also determined by its ability to form hyphae. The hyphal forms help it to adhere to the cells and lead to invasion by penetration of the epithelial lining [42]. The hyphal forms by themselves are unable to bring out cell damage and initiate a response in the host immune system [43]. Recent studies have indicated that a toxin, Candidalysin, secreted by *C. albicans,* can help in the invasion and activation of cytokines in the host [44]. It has a cytolytic action and does so by inserting itself into the host cell membrane's lipid bilayer. Candidalysin is encoded by the *ECE1* gene and is released as a pre-protein which is cleaved by the Kex2p into eight smaller peptides [45]. ECE1-III is the active peptide known as "Candidalysin." Hyphal formation is shown to be associated with increased expression of *ECE1,* which in turn results in the production of candidalysin. This toxin consists of 31 amino acids (SIIGIIMGILGNIPQVIQIIMSIVKAFKGNK) and is secreted by the hyphae in the surrounding milieu of the host cell [46]. It interacts with the host cells by forming pore-like structures that result in calcium influx and release of lactate dehydrogenase (LDH) on the host cell membrane. Epidermal growth factor receptors (EGFR) have also been proposed to help in the internalization of candidalysin. EGFR is a membrane-bound tyrosine kinase present throughout the human body and is also elevated in epithelial dysplasias and oral squamous cell carcinomas [47]. EGFR plays a vital role in activating mitogen-activated protein kinase (MAPK) cascades, which are implicated to have a role in cellular proliferation [48].

Candidalysin also affects the host's immune system by bringing about macrophage lysis and inflammasome formation [49]. Inflammasomes are multi-protein complexes that are found to have a close association with carcinogenesis. These consist of nucleotide-binding domain, leucine-rich repeat-containing receptor (NLR), self-cleaved caspase-1 and apoptosis-associated speck-like protein containing a CARD domain (ASC) [50].

Candidalysin results in activation of Nod-like receptor family pyrin domain-containing protein 3 (NLRP3) inflammasome by upregulating K+ efflux pumps [49, 51]. NLRP3 has been shown to have diverse actions promoting as well as decreasing tumorigenesis [50, 52]. Patients with oral squamous cell carcinomas have been shown to have a high expression of NLRP3 [52]. Thus, this reiterates that there exists an association between *C. albicans* secreting Candidalysin and OSCC.

Activation of cytokines and immunomodulatory substances by Candidalysin also plays an important role in the causation of OSCC. Damage of the epithelial cells by *C. albicans* leads to the release of various pro-interleukins such as IL-1α, IL-1β, IL-6, IL-23, granulocyte-macrophage colony-stimulating factor (GM-CSF), etc. [53]. This triggers the expression of various genes, such as homeobox (HOX) genes. Expression of proteins, namely HOXB7 and collagen type IV, are also influenced by these factors [54]. All these inflammatory modulators also cause cell injury. The resulting cell injury can also lead to precancerous lesions, which can progress to neoplasms [54, 55].

Candidalysin has also been shown to stimulate the production of GM-CSF [55]. It is a 24–33 kDa glycoprotein that has been studied for its role in carcinogenesis [56]. It has been shown to have a role in the proliferation of human endothelial cells and is also used to stimulate the proliferation of various tumor cell lines [55]. Moreover, it has also been shown to be produced by various cancer cells [57]. Thus, infection or colonization by *C. albicans* can result in increased secretion of GM-CSF with the help of Candidalysin. This, in turn, may lead to cellular proliferation and activation of precancerous cells to form carcinomas.

11.2.6 Destruction of Basement Membrane

Destruction of basement membrane or stromal tissue due to various factors has also been thought to play a vital role in the causation of many cancers [58]. *C. albicans* have been known to produce various enzymes, such as metallopeptidases, that can damage the basement membrane and the surrounding tissue [59]. It can lead to the degradation of the basement membrane by destroying laminin-332 [60]. Human oral E-cadherin, a glycoprotein helping in the adhesion of keratinocytes, is also affected by *C. albicans* [61]. Thus, infection or colonization by *C. albicans* can result in the malignant transformation of cells.

11.3 Conclusion

In conclusion, fungal infections are associated with oral cancers. More studies are warranted to ascertain the causal relationship between oral cancers and Candida colonization or infection in the oral mucosa.

References

1. Stajich JE, Berbee ML, Blackwell M, Hibbett DS, James TY, Spatafora JW, Taylor JW (2009) The fungi. Curr Biol 19(18):R840–R845. https://doi.org/10.1016/j.cub.2009.07.004

2. Barrett JR (2005) Liver cancer and aflatoxin: new information from the Kenyan outbreak. Environ Health Perspect 113(12):A837–A838
3. Kew MC (2013) Aflatoxins as a cause of hepatocellular carcinoma. J Gastrointestin Liver Dis 22(3):305–310
4. Sankari SL, Gayathri K, Balachander N, Malathi L (2015) Candida in potentially malignant oral disorders. J Pharm Bioallied Sci 7(Suppl 1):S162–S164. https://doi.org/10.4103/0975-7406.155886
5. Montero PH, Patel SG (2015) Cancer of the oral cavity. Surg Oncol Clin N Am 24(3):491–508. https://doi.org/10.1016/j.soc.2015.03.006
6. Gholizadeh P, Eslami H, Yousefi M, Asgharzadeh M, Aghazadeh M, Kafil HS (2016) Role of oral microbiome on oral cancers, a review. Biomed Pharmacother 84:552–558. https://doi.org/10.1016/j.biopha.2016.09.082
7. Hamid GA, Baom N (2017) Tobacco and betel quid in development of oral cancer. J Cancer Prev Curr Res 7(1):00223. https://doi.org/10.15406/jcpcr.2017.07.00223
8. Ranganathan K, Kavitha L (2019) Oral epithelial dysplasia: classifications and clinical relevance in risk assessment of oral potentially malignant disorders. J Oral Maxillofac Pathol 23(1):19–27. https://doi.org/10.4103/jomfp.JOMFP_13_19
9. Kany S, Vollrath JT, Relja B (2019) Cytokines in inflammatory disease. Int J Mol Sci 20(23). https://doi.org/10.3390/ijms20236008
10. Scully C, el-Kabir M, Samaranayake LP (1994) Candida and oral candidosis: a review. Crit Rev Oral Biol Med 5(2):125–157. https://doi.org/10.1177/10454411940050020101
11. Mun M, Yap T, Alnuaimi AD, Adams GG, McCullough MJ (2016) Oral candidal carriage in asymptomatic patients. Aust Dent J 61(2):190–195. https://doi.org/10.1111/adj.12335
12. Jain M, Shah R, Chandolia B, Mathur A, Chauhan Y, Chawda J, Mosby S, Bhagalia S (2016) The oral carriage of Candida in oral cancer patients of Indian origin undergoing radiotherapy and/or chemotherapy. J Clin Diagn Res 10(2):ZC17–ZC20. https://doi.org/10.7860/JCDR/2016/15702.7180
13. Warnakulasuriya S, Johnson NW, van der Waal I (2007) Nomenclature and classification of potentially malignant disorders of the oral mucosa. J Oral Pathol Med 36(10):575–580. https://doi.org/10.1111/j.1600-0714.2007.00582.x
14. Krogh P, Holmstrup P, Thorn JJ, Vedtofte P, Pindborg JJ (1987) Yeast species and biotypes associated with oral leukoplakia and lichen planus. Oral Surg Oral Med Oral Pathol 63(1):48–54. https://doi.org/10.1016/0030-4220(87)90339-2
15. Shukla K, Vun I, Lov I, Laparidis G, McCamley C, Ariyawardana A (2019) Role of Candida infection in the malignant transformation of oral leukoplakia: a systematic review of observational studies. Transl Res Oral Oncol:4:2057178X19828229. https://doi.org/10.1177/2057178X19828229
16. Silverman S Jr, Gorsky M, Lozada F (1984) Oral leukoplakia and malignant transformation. A follow-up study of 257 patients. Cancer 53(3):563–568. https://doi.org/10.1002/1097-0142(19840201)53:3<563::aid-cncr2820530332>3.0.co;2-f
17. Abdulrahim MH, McManus BA, Flint SR, Coleman DC (2013) Genotyping Candida albicans from Candida leukoplakia and non-Candida leukoplakia shows no enrichment of multilocus sequence typing clades but enrichment of ABC genotype C in Candida leukoplakia. PLoS One 8(9):e73738. https://doi.org/10.1371/journal.pone.0073738
18. Banoczy J, Sugar L (1975) Progressive and regressive changes in Hungarian oral leukoplakias in the course of longitudinal studies. Community Dent Oral Epidemiol 3(4):194–197. https://doi.org/10.1111/j.1600-0528.1975.tb00307.x
19. Hebbar PB, Pai A, D S (2013) Mycological and histological associations of Candida in oral mucosal lesions. J Oral Sci 55 (2):157–160. doi:https://doi.org/10.2334/josnusd.55.157
20. Shah N, Ray JG, Kundu S, Sardana D (2017) Surgical management of chronic hyperplastic candidiasis refractory to systemic antifungal treatment. J Lab Physicians 9(2):136–139. https://doi.org/10.4103/0974-2727.199622

21. Cawson RA, Lehner T (1968) Chronic hyperplastic candidiasis—candidal leukoplakia. Br J Dermatol 80(1):9–16. https://doi.org/10.1111/j.1365-2133.1968.tb11899.x
22. Eyre J, Nally FF (1971) Oral candidosis and carcinoma. Br J Dermatol 85(1):73–75. https://doi.org/10.1111/j.1365-2133.1971.tb07183.x
23. Sitheeque MA, Samaranayake LP (2003) Chronic hyperplastic candidosis/candidiasis (candidal leukoplakia). Crit Rev Oral Biol Med 14(4):253–267. https://doi.org/10.1177/154411130301400403
24. Krogh P, Hald B, Holmstrup P (1987) Possible mycological etiology of oral mucosal cancer: catalytic potential of infecting Candida albicans and other yeasts in production of N-nitrosobenzylmethylamine. Carcinogenesis 8(10):1543–1548. https://doi.org/10.1093/carcin/8.10.1543
25. McCullough M, Jaber M, Barrett AW, Bain L, Speight PM, Porter SR (2002) Oral yeast carriage correlates with presence of oral epithelial dysplasia. Oral Oncol 38(4):391–393. https://doi.org/10.1016/s1368-8375(01)00079-3
26. Alnuaimi AD, Wiesenfeld D, O'Brien-Simpson NM, Reynolds EC, McCullough MJ (2015) Oral Candida colonization in oral cancer patients and its relationship with traditional risk factors of oral cancer: a matched case-control study. Oral Oncol 51(2):139–145. https://doi.org/10.1016/j.oraloncology.2014.11.008
27. Patil S (2020) Analyzing the association between Candida prevalence, species specificity, and oral squamous cell carcinoma: a systematic review and meta-analysis—Candida and OSCC. Appl Sci 10(3):1099
28. Sankari SL, Mahalakshmi K, Kumar VN (2020) A comparative study of Candida species diversity among patients with oral squamous cell carcinoma and oral potentially malignant disorders. BMC Res Notes 13(1):488. https://doi.org/10.1186/s13104-020-05336-3
29. Chattopadhyay I, Verma M, Panda M (2019) Role of oral microbiome signatures in diagnosis and prognosis of oral cancer. Technol Cancer Res Treat 18:1533033819867354. https://doi.org/10.1177/1533033819867354
30. Perera M, Al-Hebshi NN, Perera I, Ipe D, Ulett GC, Speicher DJ, Chen T, Johnson NW (2017) A dysbiotic mycobiome dominated by Candida albicans is identified within oral squamous-cell carcinomas. J Oral Microbiol 9(1):1385369. https://doi.org/10.1080/20002297.2017.1385369
31. Karpinski TM (2019) Role of oral microbiota in cancer development. Microorganisms 7(1):20. https://doi.org/10.3390/microorganisms7010020
32. Shay E, Sangwan N, Padmanabhan R, Lundy S, Burkey B, Eng C (2020) Bacteriome and mycobiome and bacteriome-mycobiome interactions in head and neck squamous cell carcinoma. Oncotarget 11(25):2375–2386. https://doi.org/10.18632/oncotarget.27629
33. Filler SG, Sheppard DC (2006) Fungal invasion of normally non-phagocytic host cells. PLoS Pathog 2(12):e129. https://doi.org/10.1371/journal.ppat.0020129
34. Phan QT, Myers CL, Fu Y, Sheppard DC, Yeaman MR, Welch WH, Ibrahim AS, Edwards JE Jr, Filler SG (2007) Als3 is a Candida albicans invasin that binds to cadherins and induces endocytosis by host cells. PLoS Biol 5(3):e64. https://doi.org/10.1371/journal.pbio.0050064
35. Moyes DL, Richardson JP, Naglik JR (2015) Candida albicans-epithelial interactions and pathogenicity mechanisms: scratching the surface. Virulence 6(4):338–346. https://doi.org/10.1080/21505594.2015.1012981
36. Dwivedi PP, Mallya S, Dongari-Bagtzoglou A (2009) A novel immunocompetent murine model for Candida albicans-promoted oral epithelial dysplasia. Med Mycol 47(2):157–167. https://doi.org/10.1080/13693780802165797
37. Mohd Bakri M, Mohd Hussaini H, Rachel Holmes A, David Cannon R, Mary Rich A (2010) Revisiting the association between candidal infection and carcinoma, particularly oral squamous cell carcinoma. J Oral Microbiol 2. https://doi.org/10.3402/jom.v2i0.5780
38. Davidson BJ, Hsu TC, Schantz SP (1993) The genetics of tobacco-induced malignancy. Arch Otolaryngol Head Neck Surg 119(11):1198–1205. https://doi.org/10.1001/archotol.1993.01880230040007

39. Zakhari S (2006) Overview: how is alcohol metabolized by the body? Alcohol Res Health 29(4):245–254
40. Homann N, Tillonen J, Meurman JH, Rintamaki H, Lindqvist C, Rautio M, Jousimies-Somer H, Salaspuro M (2000) Increased salivary acetaldehyde levels in heavy drinkers and smokers: a microbiological approach to oral cavity cancer. Carcinogenesis 21(4):663–668. https://doi.org/10.1093/carcin/21.4.663
41. Bakri MM, Rich AM, Cannon RD, Holmes AR (2015) In vitro expression of Candida albicans alcohol dehydrogenase genes involved in acetaldehyde metabolism. Mol Oral Microbiol 30(1): 27–38. https://doi.org/10.1111/omi.12064
42. Arkowitz R, Bassilana M (2019) Recent advances in understanding Candida albicans hyphal growth [version 1; peer review: 4 approved]. F1000Res 8(700). https://doi.org/10.12688/f1000research.18546.1
43. Moyes DL, Wilson D, Richardson JP, Mogavero S, Tang SX, Wernecke J, Hofs S, Gratacap RL, Robbins J, Runglall M, Murciano C, Blagojevic M, Thavaraj S, Forster TM, Hebecker B, Kasper L, Vizcay G, Iancu SI, Kichik N, Hader A, Kurzai O, Luo T, Kruger T, Kniemeyer O, Cota E, Bader O, Wheeler RT, Gutsmann T, Hube B, Naglik JR (2016) Candidalysin is a fungal peptide toxin critical for mucosal infection. Nature 532(7597):64–68. https://doi.org/10.1038/nature17625
44. Wilson D, Naglik JR, Hube B (2016) The missing link between Candida albicans hyphal morphogenesis and host cell damage. PLoS Pathog 12(10):e1005867. https://doi.org/10.1371/journal.ppat.1005867
45. Naglik JR, Gaffen SL, Hube B (2019) Candidalysin: discovery and function in Candida albicans infections. Curr Opin Microbiol 52:100–109. https://doi.org/10.1016/j.mib.2019.06.002
46. Konig A, Hube B, Kasper L (2020) The dual function of the fungal toxin candidalysin during Candida albicans-macrophage interaction and virulence. Toxins (Basel) 12(8):doi:10.3390/toxins12080469
47. Ribeiro DC, Gleber-Netto FO, Sousa SF, Bernardes VD, Guimaraes-Abreu MH, Aguiar MC (2012) Immunohistochemical expression of EGFR in oral leukoplakia: association with clinicopathological features and cellular proliferation. Med Oral Patol Oral Cir Bucal 17(5):e739–e744. https://doi.org/10.4317/medoral.17950
48. Zhang W, Liu HT (2002) MAPK signal pathways in the regulation of cell proliferation in mammalian cells. Cell Res 12(1):9–18. https://doi.org/10.1038/sj.cr.7290105
49. Kasper L, Konig A, Koenig PA, Gresnigt MS, Westman J, Drummond RA, Lionakis MS, Gross O, Ruland J, Naglik JR, Hube B (2018) The fungal peptide toxin Candidalysin activates the NLRP3 inflammasome and causes cytolysis in mononuclear phagocytes. Nat Commun 9(1): 4260. https://doi.org/10.1038/s41467-018-06607-1
50. Thi HTH, Hong S (2017) Inflammasome as a therapeutic target for cancer prevention and treatment. J Cancer Prev 22(2):62–73. https://doi.org/10.15430/JCP.2017.22.2.62
51. Munoz-Planillo R, Kuffa P, Martinez-Colon G, Smith BL, Rajendiran TM, Nunez G (2013) K (+) efflux is the common trigger of NLRP3 inflammasome activation by bacterial toxins and particulate matter. Immunity 38(6):1142–1153. https://doi.org/10.1016/j.immuni.2013.05.016
52. Wang H, Luo Q, Feng X, Zhang R, Li J, Chen F (2018) NLRP3 promotes tumor growth and metastasis in human oral squamous cell carcinoma. BMC Cancer 18(1):500. https://doi.org/10.1186/s12885-018-4403-9
53. Li X, Leonardi I, Iliev ID (2017) Candidalysin sets off the innate alarm. Sci Immunol 2(17). https://doi.org/10.1126/sciimmunol.aao5703
54. Rivera C, Venegas B (2014) Histological and molecular aspects of oral squamous cell carcinoma (review). Oncol Lett 8(1):7–11. https://doi.org/10.3892/ol.2014.2103
55. Engku Nasrullah Satiman EAF, Ahmad H, Ramzi AB, Abdul Wahab R, Kaderi MA, Wan Harun WHA, Dashper S, McCullough M, Arzmi MH (2020) The role of Candida albicans candidalysin ECE1 gene in oral carcinogenesis. J Oral Pathol Med 49(9):835–841. https://doi.org/10.1111/jop.13014

56. Hong IS (2016) Stimulatory versus suppressive effects of GM-CSF on tumor progression in multiple cancer types. Exp Mol Med 48(7):e242. https://doi.org/10.1038/emm.2016.64
57. Nebiker CA, Han J, Eppenberger-Castori S, Iezzi G, Hirt C, Amicarella F, Cremonesi E, Huber X, Padovan E, Angrisani B, Droeser RA, Rosso R, Bolli M, Oertli D, von Holzen U, Adamina M, Muraro MG, Mengus C, Zajac P, Sconocchia G, Zuber M, Tornillo L, Terracciano L, Spagnoli GC (2014) GM-CSF production by tumor cells is associated with improved survival in colorectal cancer. Clin Cancer Res 20(12):3094–3106. https://doi.org/10.1158/1078-0432.CCR-13-2774
58. Natarajan E, Saeb M, Crum CP, Woo SB, McKee PH, Rheinwald JG (2003) Co-expression of p16(INK4A) and laminin 5 gamma2 by microinvasive and superficial squamous cell carcinomas in vivo and by migrating wound and senescent keratinocytes in culture. Am J Pathol 163(2):477–491. https://doi.org/10.1016/s0002-9440(10)63677-2
59. Rodier MH, el Moudni B, Kauffmann-Lacroix C, Daniault G, Jacquemin JL (1999) A Candida albicans metallopeptidase degrades constitutive proteins of extracellular matrix. FEMS Microbiol Lett 177(2):205–210. https://doi.org/10.1111/j.1574-6968.1999.tb13733.x
60. Parnanen P, Kari K, Virtanen I, Sorsa T, Meurman JH (2008) Human laminin-332 degradation by Candida proteinases. J Oral Pathol Med 37(6):329–335. https://doi.org/10.1111/j.1600-0714.2008.00638.x
61. Parnanen P, Meurman JH, Samaranayake L, Virtanen I (2010) Human oral keratinocyte E-cadherin degradation by Candida albicans and Candida glabrata. J Oral Pathol Med 39(3):275–278. https://doi.org/10.1111/j.1600-0714.2009.00866.x

Microbial "OMICS" in Oral Cancer

12

Shaheerah Khan, Rohita Sinha, and Anshuman Dixit

Abstract

Oral squamous cell carcinoma (OSCC) is the leading cause of mortality in head and neck cancer. Unhealthy habits such as chewing tobacco, smoking, drinking, poor oral hygiene and diet are responsible for chronic inflammation and infections have been recognized as predisposing factors for oral carcinogenesis. Various bacterial species and viruses are known to be associated with oral cancer. The metabolic by-products of the bacterial flora induce permanent genetic alterations in epithelial cells of the host that may drive carcinogenesis. Recent advancements in "omics" technologies have been useful in identifying oral cancer-related microbiome, their genomes, virulence properties, and their interaction with host immunity. To understand the journey of a microbial community from supporting a healthy environment to transitioning into a diseased state, multi-omics approaches are being utilized. These approaches not only provide insight into the microbiome but they can also provide detailed information at the DNA (genome), RNA (transcriptome), protein (proteome), and metabolite (metabolome) level as well. The differential genomic and transcriptomic signatures would lead to the functional identity of the microbial community in a particular environment leading to alterations in the host biological process and pathways. The OMICS-based approaches are powerful tools available to researchers to understand the physiological and pathological roles of microbes in a host.

S. Khan · A. Dixit (✉)
Institute of Life Science, Bhubaneswar, Odisha, India
e-mail: khshaheerah@gmail.com; anshumandixit@gmail.com

R. Sinha
University of Nebraska—Lincoln, Lincoln, NE, USA
e-mail: rohita.sinha@unl.edu

S. Routray (ed.), *Microbes and Oral Squamous Cell Carcinoma*,
https://doi.org/10.1007/978-981-19-0592-6_12

Keywords

Multi-omics · Microbes · Oral squamous cell carcinoma

12.1 Introduction

The high-throughput sequencing technologies have transformed life sciences research. The "omics" suffixed to a molecular term, like genomics, proteomics, metabolomics, etc. implies a comprehensive evaluation of a set of molecules (http://omics.org/). The current developments in the high-throughput technologies in omics research have led to a better understanding of the molecules and processes involved in human health and diseases. They have resulted in the development of various diagnostic, prognostic, and therapeutic strategies. Of late the research into microbial omics has led to a completely new understanding of various diseases. These omics technologies, such as metabolomics and transcriptomics, are now commonly used in biological research and are immensely contributing to our understanding of role of microbes in disease processes.

12.1.1 Human Oral Microbiome

Microbes (bacteria, fungi, viruses, protozoa) colonizing different niches of the oral cavity are called the oral microbiome. The oral microbiome is considered the most diverse in the human body after the gut [1]. Around 600–700 bacterial phylotypes have been detected in the oral cavity [2]. Vaginal birth plays an important role in predisposing the bacterial community that forms a stable ecosystem in the mouth of the infant. Therefore human oral microbiome tends to change at the micro-level (inter-individual differences) while remaining resilient at the species (human) level [3, 4].

The oral microbiome is an essential part of the oral cavity that acts as a buffer, preventing the changes there, which is important to maintain a healthy environment. The oral cavity, with a temperature of 37 °C and a pH of 6.8–7.2, provides an ideal environment for the microbiome to thrive. A change in the environmental conditions increases the potential of pathogens to create oral diseases [5]. The human oral microbiome can be broadly divided into two types:

1. *Core microbiome,* comprising predominant species existing at different sites, in a healthy condition [6, 7].
2. *Variable microbiome,* which has evolved in response to different lifestyles, food habits, etc. [6].

The core microbiome is shared among all individuals and is comprised of the predominant species that exist under healthy conditions at different sites of the body.

The oral microbiome is a complex cocktail of different microbes that not only differs in its composition (bacteria, fungi, viruses, archaea, and protozoa) but is also specific to the subsites. The human oral cavity is divided into different subsites, including the gingiva, tongue, cheeks, hard and soft palates, hard tissues (teeth), exocrine gland tissue (major and minor salivary glands), supra-gingival plaque of tooth surfaces, subgingival plaque, maxillary anterior vestibule, and tonsils, all of which are colonized by distinctive predominant microbiota with unique features. The composition of the flora in all the subsites in healthy and diseased individuals mostly belongs to phyla Actinobacteria, Bacteroidetes, Firmicutes, Proteobacteria, Spirochaetes, Synergistetes, and Tenericutes, along with viruses, fungi, protozoa, mycoplasma, and archaea.

12.1.1.1 Composition of Oral Microbiome

12.1.1.1.1 Bacteria

Human oral microbiome mostly consists of bacteria, in comparison to other microorganisms, hence the term oral bacteriome came into being. A huge collection of bacterial morphs in dental plaque was shown through the microscope, first by Leeuwenhoek. Other scientists like Koch and Pasteur showed the characteristic of the bacteria, like its cultivation through bacterial cell cultures. There are still many bacteria in the oral cavity that are yet to be cultivated and studied. Therefore, culture-independent methods like the 16S RNA sequencing opened new avenues for the detection of the bacteriome in an oral microbiome. The bacteriome of the oral cavity of a healthy individual is dominated by the genus *Actinomyces* (facultative anaerobe), *Capnocytophaga, Eikenella, Eubacteria, Fusobacterium* (obligate anaerobes), *Haemophilus, Lactobacterium, Leptotrichia, Neisseria, Porphyromonas, Prevotella, Propionibacterium, Peptostreptococcus, Streptococcus, Staphylococcus, Veillonella* (obligate anaerobes), and *Treponema* [8, 9]. There are bacterial communities with an affinity for different subsites in the oral cavity such as periodontal pockets, teeth, tongue, etc. The tongue shows the highest diversity of microbiota among them. Microbes in the tongue like *Veillonella atypica, Porphyromonas gingivalis, Selenomonas subspecies, Aggregatibacter, actinomycetemcomitans, Prevotella intermedia, Capnocytophaga, Enterococcus faecalis,* and *Eikenella corrodens* facilitate colonization of bacteria in other regions of the oral cavity through saliva. The composition of other sites in the oral cavity like oropharynx are *Streptococcus pyogenes, Streptococcus pneumoniae, Haemophilus influenza, Haemophilus parainfluenzae, Streptococcus mutans, Streptococcus salivarius, Streptococcus anginosus;* the surface of the teeth contains *S. mutans, Actinomyces, Eubacterium, Peptostreptococcus*; Tonsil has *Streptococcus viridans, H. influenza, Neisseria species, Staphylococcus*; Gingiva has *Fusobacterium, Prevotella, Porphyromonas, Streptococcus mitis, Streptococcus sanguinis, Propionibacterium acnes Leptotrichia buccalis, Actinomyces odontolyticus, Veillonella parvula*; and dental plaque harbors *Actinomyces, Rothia, Microbacterium, Mycobacterium Propionibacterium,* and *Corynebacterium Bifidobacterium* [8–10].

The presence of bacteria (aerobic and anaerobic) helps maintain a healthy environment, by creating biofilms that do not allow a change in the environment, but certain circumstances change the composition of the microbiome, disturbing the balance, creating dysbiosis, that finally leads to diseased conditions [11]. Dysbiosis in turn leads to chronic inflammatory conditions. Oral commensal bacteria play important roles in the development of oral diseases such as periodontitis [12]. They are also known to cause cancer. Two specific bacterial species *Fusobacterium nucleatum* (a gram-negative commensal) and *Porphyromonas gingivalis* are shown in many studies to be involved in cancer development. *P. gingivalis* has been shown to suppress apoptosis [13, 14], suggesting that the anti-apoptotic effect of *P. gingivalis* may be involved in oral carcinogenesis. Similarly, *F. nucleatum* activates cell signaling that facilitates immune evasion, leading to carcinogenesis [15].

12.1.1.1.2 Virus

A virome is a collection of all the viruses that are present in the human body. They influence human health and are associated with various diseases. The human virome is individual-specific, and sometimes includes certain uncharacterized bacteriophages, which may play a defensive role against the pathogens [16, 17]. Most microbes designated as Class 1 carcinogens by the International Agency for Research on Cancer (IARC) are viruses. There are certain viruses that are involved in oral, oro-pharyngeal, sino-nasal, and laryngeal carcinomas, such as the human papilloma virus (HPV). HPV has also been shown to be involved in premalignant conditions like leukoplakia and erythroplakia [18]. Similarly, Epstein–Barr *virus* (EBV) is known to cause salivary gland tumors [19]. There are certain blood-borne viruses, like Hepatitis virus and HIV that are known to be present in the mouth as well. Herpes virus has been known to cause gingivostomatitis and cold sores [20] and other oral conditions, like papilloma [21], condyloma and focal epithelial hyperplasia [22].

12.1.1.1.3 Protozoa

The protozoan species such as *Trichomonas tenax* and *Entamoeba gingivalis* are found in a normal oral microbiome. At present, though regarded harmless, they were once considered to be oral pathogens as they were found to increase in numbers in the oral cavity of patients with poor hygiene and gingival disease [23]. Protozoan species like *Toxoplasma gondii, Cryptosporidium parvum,* and *Trichomonas vaginalis* are known to cause cancer like gliomas, astrocytomas, colorectal cancer, cervical cancer, etc., but there has not been much study to come across the protozoans found in the oral cavity, that helps in oral carcinogenesis.

12.1.1.1.4 Fungi

The mycobiome of the oral microbiome also participates in either maintaining the normal microbiome or in disrupting one. Candida sp. is the most commonly found fungi in the oral cavity and causes a wide range of infections [24]. Studies have shown that a healthy oral mycobiome consists of around more than 80 genera of fungi, including *Cladosporium, Aureobasidium, Saccharomycetales, Aspergillus, Fusarium, Cryptococcus,* and *Candida* being the most predominant [25].

The role of oral mycobiome has been studied with respect to premalignant oral conditions and oral squamous cell carcinoma (OSCC). These studies used nitrosation potential, Perodic Acid-Schiff (PAS) staining, and cultivation techniques to detect and isolate fungi from oral swabs, biopsies, lesions, dysplastic and cancerous tissues [26–28]. The applications of next-generation sequencing (NGS) have opened more opportunities to detect numerous other species, which remained undetected by cultivation techniques [29]. The use of novel technologies like the cultivation technique in combination with matrix-assisted laser desorption ionization time-of-flight mass spectrometry (MALDI-TOF MS) helps in a more accurate and detailed detection of oral cancer associated mycobiome with *Rhodotorula, Saccharomyces,* and *Kloeckera* in addition to Candida species [30].

12.1.1.1.5 Archaea

The Archaeal community is found at a minimal amount in the oral microbiome, restricted to a small number of species, most of which are methanogens (methane-producing archaea). They can be found in healthy as well as diseased subjects, with a significant difference in their number. Most abundantly found methanogens are *Methanobrevibacter oralis, Methanobrevibacter phylotypes, Methanobacterium curvum/congolense,* and *Methanosarcina mazeii* [31, 32]. Molecular cultivation and cultivation-independent (16S rRNA sequencing) methods have been used to detect these methanogens, most of which have been collected from periodontal plaque, as they exclusively reside at sites of poly-microbial anaerobic biofilms in the oral cavity (e.g., periodontal pockets) [33]. The functional gene *mcrA* for the enzyme Methyl-Coenzyme M reductase involved in methanogenesis is detected through PCR technique that validates its predominance. A study has also shown the presence of Thermoplasma-like organisms belonging to Euryarchaeota, in the oral cavity [34]. Methanogens, though present in healthy individuals, are also potent enough to cause harm to the host, because of their capability of transforming heavy metals and metalloids into toxic volatile methylated derivatives that can disrupt the niche [35]. Though there has not been much evidence to prove they can cause oral cancer, they can produce hypoxic conditions that might indirectly lead to cancer.

12.1.2 Holo-Omics

The new sequencing technologies are helping us in many ways to understand the roles that the microbiome play in the initiation and disposition of diseases. The technologies like the amplicon sequencing of 16s rRNA genes, or recent shotgun metagenomics of microbiomes have helped scientists look deep into their taxonomy and abundance. However, to fully understand the dynamics of the community, a comprehensive systems-level study consisting of the information related not only to DNA but RNA, protein, and metabolite level as well, termed as "multi-omics" is required [1]. Moreover, it is important to not only study the multiome of the microbes but also the environment it lives in, to understand the host-microbiome interaction, pathogenesis, and survival. A recent terminology that is being used with host and microbiome studies is "Holo-omics." It is the study and analysis of multiple omic levels from both the host and its associated microbiota. The term "Hologenome" refers to the combined genetic content of the host (holobiont) and its associated microbiota. The hologenome theory of evolution regarding the microbiome refers to the interactive and symbiotic existence of both the host and microbe in an environment throughout the process of evolution [2]. A Holo-omics study that consists of the metagenome, metatranscriptome, metaproteome, and meta-metabolome [3] can help decipher the molecular markers and the interplay between the host and microbial genomes, especially in cancer. The role microbiome plays in oral squamous cell carcinoma (OSCC) remains unexplored, yet there have been some reports that put forward an idea of the micro-changes at RNA, protein, and metabolite level that changes the downstream biological processes and pathways.

12.1.3 Meta-Transcriptomics

Meta-transcriptomics is the study of bacterial transcriptomes [4]. It refers to the study of entire pool of mRNA in a microbial population of a given environment [5]. The advantage of studying the gene expression of the bacterial genome at the transcript level helps understand the functional aspect of microbial communities clearly [6]. Meta-transcriptomics helps understand the link between environment and microbial population and how they respond to changes in their environment.

A study to explore the meta-transcriptomics of the oral microbiome in healthy and 4-NQO (4-nitroquinoline 1-oxide)-induced tumor mice showed a significant difference pertaining to the population of microbes and their genes expression. The abundance of Genus Parabacteroides, a gram-negative, anaerobic bacteria increased, while Genus Cornybacterium, a gram-positive, aerobic bacteria decreased in the oral cancer group (4-NQO-induced tumor mice). These genes when analyzed further for their functional annotation revealed overrepresented activities like nitrogen transport, response to radiation, and stress (oxidative stress), proteolysis, cell wall biosynthesis, N-acetylneuraminate catabolism, amino acid metabolism, actin filament depolymerization, and Wnt signaling pathway. In addition, there was an increase in the immune response including T cells and myeloid cells in the areas treated with 4-NQO [7].

The other studies include the study of metatranscriptome in oral diseases, like dental caries, gingival plaques, and periodontitis, to have a better view of the population and gene expression that may underlie pathological functions. The abundance of *Streptococcus mutans* in dental caries along with genera Fusobacterium, Actinomyces, Porphyromonas, Haemophilus, etc., showed a dysregulation of phosphotransferase in the beta-glucoside metabolism that is seen to help in colonization of the microbial community [8, 9], fructose, and mannose metabolism, disease pathways that convert sorbitol (a carbon source) to fructose 6-phosphate [10, 11]. The metatrascriptome of subgingival plaque in periodontitis revealed the "red complex" species comprising of *Porphyromonas gingivavlis*, *Treponema denticola*, and *Tannerella forsythia* that are responsible for chronic periodontitis [11]. Their presence not only showed high expression of metalloproteases and also genes that were associated with invasion [12], but also showed overrepresented biological processes, such as flagellar motility, iron acquisition, transport and secretion, antimicrobial resistance, histidine catabolism, lipid and peptidoglycan synthesis, protein kinase signaling pathway, metal ion transport, and beta-lactam degradation, and underrepresented processes like potassium transport and polysaccharide biosynthesis [13–15]. Other studies comparing tumor and adjacent normal buccal cavity sites also reveal the presence of the carcinogenic microbes with over- and under-represented biological processes like the increased haemolysin activity on host erythrocytes, and lactate and phosphate transport, while the underrepresented activity included nucleotide excision repair mechanism. The overrepresented molecular functions included ion binding, as well as NO-reductase, lactoylglutathione lyase, tryptophanase, metalloexopeptidase, urease, and glutamate dehydrogenase (GDH) activity [16].

12.1.4 Metaproteomics

Metaproteomics is a study of the presence and abundance of proteins. It can provide insights into the phenotypes in each microbial community at a molecular level. It was previously used as a tool for the large-scale analysis of proteins in microbiomes [17]. Studying the metaproteome of a microbiome gives an overall view of the metabolism, community dynamics, and its composition as well. It very well defines the interactions between the host and microbes [18, 19]. Metaproteomics also involves differential gene expression analysis, as certain genes are important in shaping the structure of the community and environment [20, 21]. Metaproteomics have also been used to study transposon mutagenesis in different species with the ultimate aim to develop microbiota-conditioned foods, which showed a significant difference in a community when the dietary conditions were changed [22]. Metaproteomics have been used to study the host-microbiome interactions, which further helps the researchers to understand the etiology of the disease [23–26].

A study involving the proteomics analysis of saliva from the oral cancer patients identified proteins involved in OSCC progression were *STAT3* [27, 28], *RIPK2*, [29], IKBKB [30], and *PRDX3* [31, 32], that activates NF-κB; OSCC tumor markers such as *SCCA1* and *SCCA2*, [33], *CD82* [34], *SLC3A2* [35], *LY6D* [36], *CD9* [37], and *TACSTD2* [38].

A deep metaproteomic study of the human saliva from healthy individuals found the relation of the microbial community harboring genera *Streptococcus*, *Rothia*, *Actinomyces*, *Prevotella*, *Neisseria*, *Veilonella*, *Lactobacillus*, *Selenomonas*, *Pseudomonas*, *Staphylococcus*, *Campylobacter,* and TM7 (phylum Saccharibacteria), with the carbohydrate, amino acid and energy metabolism [39]. To understand the shift in the metaproteome from normal to a diseased state, another study showed a metaproteomic study where the proteins responsible for the development of oral cancer were identified in the saliva. It found the abundance of genera *Streptococcus*, *Neisseria*, and *Haemophilus*, mostly. Further analysis showed their involvement in protein translation and glycolysis. A similar study also found and validated previous findings where pathways related to carbohydrate, amino acid, and energy metabolism, with processes like signal transduction and membrane transport were highly represented in the presence of those microbes, in oral cancer patients [40, 41]. A study dedicated to analyzing the metaproteome of the saliva of oral cancer patients, found the largest number of peptides designated to phylum Firmicutes and half of that to genus *Streptococcus*, followed by phylum Proteobacteria. A very less percentage of peptides were assigned to double-stranded DNA viruses and phylum Archae. Microbes found specifically in the nasopharynx were *Haemophilus influenza*, *Neisseria meningitides*, and *Streptococcus pneumoniae*. The study of these peptides found links to many protein pathways such as translation and the peptides associated with the process were tRNA synthetases, elongation factors (Tu, G), and 50S ribosomal proteins (L7/L12 and L5); carbohydrate transport and metabolism associated with proteins enolase and phosphoglycerate kinase; amino acid transport and metabolism that included different proteins assigned to its relevant pathways, such as homocysteine transmethylase in methionine biosynthesis, ketol acid reductoisomerase in valine/leucine/isoleucine biosynthesis, ornithine transcarbamylase in arginine biosynthesis/degradation, and NAD-specific glutamate dehydrogenase in glutamate degradation. The protein phosphoenolpyruvate carboxykinase in gluconeogenesis and pyruvate oxidase and formate dehydrogenase in fermentation were a few proteins of the most predominant pathways [40]. The metaproteomics study has been integrated with other "meta-omics" approaches, such as metagenomics and metatranscriptomics [42–44], to have a holistic approach towards the understanding of the disease.

12.1.5 Meta-Metabolomics

The living cells are undergoing various biochemical processes and metabolic activities daily that produces many metabolites constituting a "metabolome." The study that identifies and quantifies these metabolites is known as "metabolomics" [45]. Since microbes play an important role in altering and balancing the biological processes in the host cells, hence the term "microbial metabolomics" or "microbial meta-metabolomics" has emerged, which provides a whole picture of host-microbiota metabolism and interactions. The microbial meta-metabolome is the combination of the abundance of microbes (metagenomics) and their corresponding gene expressions (metatranscriptomics), that helps understand its effects on human health [46]. Metabolomics could be used to assess the metabolite profile either globally (untargeted approach) or specifically (targeted approach) [47]. Since bacteria produce various metabolites, these products might contribute towards the initiation of oral diseases [48]. Previous studies have demonstrated the role of lactic acid, a product of microbial sugar metabolism, in dental caries [49], while periodontal disease is initiated by the short-chain fatty acids and sulfur compounds generated through the protein and amino acids metabolism [50–52]. Cancer cells differ from normal cells based upon their metabolic profile. A common characteristic of cancer cells produces lactate as a product of aerobic glycolysis, a common characteristic often happening and referred as the "Warburg effect" [53]. A disordered metabolism contributes to cancer progression. There have been studies that relate the metabolites and the microbes in causing the Warburg effect in the OSCC tissues [54, 55]. Therefore, the knowledge of the overall metabolic profile of microbe and host cells would be of immense help to find the functional changes leading to malignancy.

To study the normal or diseased state of the oral microbial community, saliva is very crucial for such roles. Salivary metabolites have been used for risk stratification in head and neck cancer. For the metabolomics and meta-metabolomics study, two types of saliva are considered, which are the stimulated and unstimulated whole saliva, both differing in their chemical composition [56–58]. Studies have shown the role of the microbiome in the tumor initiation and progression. The microbial community in the gut has a role to play in estrogen metabolism which causes a spike in estrogen levels, giving a rise in the chances of breast and uterine cancer [59]. The metabolites such as ellagic acid and urolithins produced by the microbes help in the progression of cancer [60]. The meta-metabolomic study on the saliva of oral squamous cell carcinoma patients showed the presence of those metabolites with an increased population of *Capnocytophaga gingivalis*, *Capnocytophaga ochracea*, *Eubacterium sabureum*, *Leptotrichia buccalis*, and *Streptococcus mitis* [61, 62].

A study to check the difference in the metabolite profile of normal and cancer patients was conducted, where five metabolites such as valine, phenylalanine, n-eicosanoic acid, γ-aminobutyric acid and lactic acid, among 41 were found to be potent enough to differentiate between a normal and cancerous state [63]. Another study on saliva showed a significantly increased level of polyamine in oral cancer [64]. High polyamines levels increase cell proliferation, invasion, metastasis, and decreased apoptosis [65]. The levels of polyamines like putrescine and cadaverine decline in cancer patients undergoing radiotherapy, which also shows the important role of polyamine in cancer progression. Another study showed an increase in the level of glutathione in the saliva of patients suffering from oral cancer [66]. The higher levels of choline, phosphocholine, and lactate, which are shown to increase cell proliferation and lower levels of creatine and PUFA, shown to increase energy consumption by the tumor, were found significantly in OSCC patients [67]. A tissue level study focused on different sites in the head and neck squamous cell carcinoma, including lip, tongue, oral cavity, and larynx showed almost the same results with an increase in the levels of glutamine, glutamate, glutathione, alanine, phenylalanine, creatine, choline, taurine, tyrosine, lactate, and decreased triglycerides [68]. The metabolites associated with oral cancer are related to dysregulated metabolic pathways, like glycolysis, TCA cycle, energy metabolism, oxidative phosphorylation, and anti-oxidant mechanisms [69].

Another study showed the relation of bacterial genus Streptococcus, Prevotella, and Veillonella with an increased production of the human proteins such as GAPDH,

CATG, HRGP, BPIFA1, BPIFA2, S100A9, S100A8, NGAL, LTF, IGHA1, and ELNE in saliva of oral cancer patients [70].

Studies have utilized the meta-metabolomics in drug discovery where the microbiome can be modulated to get the best outcomes of drugs or to minimize the side effect of the drugs, shedding light on the "therapeutic metabolomics" [71, 72].

12.2 Conclusion

The "Holo-omics" study of the oral microbiome gives an overall idea from genes to the function. It is important to have a multi-omics study rather than just taxonomic classifications, to understand the mechanism of action of the microbial community at work. The differential gene expression to metabolite production and a change in it can be studied very well with the help of "omics" technologies.

References

1. Shakya M, Lo CC, Chain PSG (2019) Advances and challenges in metatranscriptomic analysis. Front Genet 10:904
2. Rosenberg E, Sharon G, Zilber-Rosenberg I (2009) The hologenome theory of evolution: a fusion of neo-Darwinism and Lamarckism. Environ Microbiol 11:2959–2962
3. Nyholm L, Koziol A, Marcos S, Botnen AB, Aizpurua O, Gopalakrishnan S et al (2020) Holo-omics: integrated host-microbiota multi-omics for basic and applied biological research. iScience 23(8):101414
4. Solbiati J, Frias-Lopez J (2018) Metatranscriptome of the oral microbiome in health and disease. J Dent Res 97(5):492–500
5. Filiatrault MJ (2011) Progress in prokaryotic transcriptomics. Curr Opin Microbiol 14(5): 579–586
6. Bashiardes S, Zilberman-Schapira G, Elinav E (2016) Use of metatranscriptomics in microbiome research. Bioinformatics Biol Insights 10:19–25
7. Stashenko P, Yost S, Choi Y, Danciu T, Chen T, Yoganathan S et al (2019) The oral mouse microbiome promotes tumorigenesis in oral squamous cell carcinoma. mSystems 4(4):e00323–e00319
8. Loo CY, Mitrakul K, Voss IB, Hughes CV, Ganeshkumar N (2003) Involvement of an inducible fructose phosphotransferase operon in Streptococcus gordonii biofilm formation. J Bacteriol 185(21):6241–6254
9. Kilic AO, Tao L, Zhang Y, Lei Y, Khammanivong A, Herzberg MC (2004) Involvement of Streptococcus gordonii beta-glucoside metabolism systems in adhesion, biofilm formation, and in vivo gene expression. J Bacteriol 186(13):4246–4253
10. Janda WM, Kuramitsu HK (1978) Production of extracellular and cell-associated glucosyltransferase activity by Streptococcus mutans during growth on various carbon sources. Infect Immun 19(1):116–122
11. Rolla G, Oppermann RV, Waaler SM, Assev S (1981) Effect of aqueous solutions of sorbitol-xylitol on plaque metabolism and on growth of Streptococcus mutans. Scand J Dent Res 89(3): 247–250
12. Duran-Pinedo AE, Chen T, Teles R, Starr JR, Wang X, Krishnan K et al (2014) Community-wide transcriptome of the oral microbiome in subjects with and without periodontitis. ISME J 8(8):1659–1672

13. Jorth P, Turner KH, Gumus P, Nizam N, Buduneli N, Whiteley M (2014) Metatranscriptomics of the human oral microbiome during health and disease. MBio 5(2):e01012–e01014
14. Yost S, Duran-Pinedo AE, Krishnan K, Frias-Lopez J (2017) Potassium is a key signal in host-microbiome dysbiosis in periodontitis. PLoS Pathog 13(6):e1006457
15. Yost S, Duran-Pinedo AE, Teles R, Krishnan K, Frias-Lopez J (2015) Functional signatures of oral dysbiosis during periodontitis progression revealed by microbial metatranscriptome analysis. Genome Med 7(1):27
16. Yost S, Stashenko P, Choi Y, Kukuruzinska M, Genco CA, Salama A et al (2018) Increased virulence of the oral microbiome in oral squamous cell carcinoma revealed by metatranscriptome analyses. Int J Oral Sci 10(4):32
17. Wilmes P, Bond PL (2004) The application of two-dimensional polyacrylamide gel electrophoresis and downstream analyses to a mixed community of prokaryotic microorganisms. Environ Microbiol 6(9):911–920
18. Salvato F, Hettich RL, Kleiner M (2021) Five key aspects of metaproteomics as a tool to understand functional interactions in host-associated microbiomes. PLoS Pathog 17(2): e1009245
19. Bhardwaj A, Kaur J, Wuest M, Wuest F (2017) In situ click chemistry generation of cyclooxygenase-2 inhibitors. Nat Commun 8(1):1
20. Kleiner M (2019) Metaproteomics: much more than measuring gene expression in microbial communities. mSystems 4(3):e00115–e00119
21. Lassek C, Burghartz M, Chaves-Moreno D, Otto A, Hentschker C, Fuchs S et al (2015) A metaproteomics approach to elucidate host and pathogen protein expression during catheter-associated urinary tract infections (CAUTIs). Mol Cell Proteomics 14(4):989–1008
22. Patnode ML, Beller ZW, Han ND, Cheng J, Peters SL, Terrapon N et al (2019) Interspecies competition impacts targeted manipulation of human gut bacteria by fiber-derived glycans. Cell 179(1):59–73. e13
23. Pathak KV, McGilvrey MI, Hu CK, Garcia-Mansfield K, Lewandoski K, Eftekhari Z et al (2020) Molecular profiling of innate immune response mechanisms in ventilator-associated pneumonia. Mol Cell Proteomics 19(10):1688–1705
24. Belda-Ferre P, Williamson J, Simon-Soro A, Artacho A, Jensen ON, Mira A (2015) The human oral metaproteome reveals potential biomarkers for caries disease. Proteomics 15(20): 3497–3507
25. Gavin PG, Mullaney JA, Loo D, Cao KL, Gottlieb PA, Hill MM et al (2018) Intestinal metaproteomics reveals host-microbiota interactions in subjects at risk for Type 1 diabetes. Diabetes Care 41(10):2178–2186
26. Tanca A, Palomba A, Fraumene C, Manghina V, Silverman M, Uzzau S (2018) Clostridial butyrate biosynthesis enzymes are significantly depleted in the gut microbiota of nonobese diabetic mice. mSphere 24:3(5)
27. Jewett A, Head C, Cacalano NA (2006) Emerging mechanisms of immunosuppression in oral cancers. J Dent Res 85(12):1061–1073
28. Shah NG, Trivedi TI, Tankshali RA, Goswami JA, Jetly DH, Kobawala TP et al (2006) Stat3 expression in oral squamous cell carcinoma: association with clinicopathological parameters and survival. Int J Biol Markers 21(3):175–183
29. Squarize CH, Castilho RM, Sriuranpong V, Pinto DS Jr, Gutkind JS (2006) Molecular cross-talk between the NFkappaB and STAT3 signaling pathways in head and neck squamous cell carcinoma. Neoplasia 8(9):733–746
30. Basseres DS, Baldwin AS (2006) Nuclear factor-kappaB and inhibitor of kappaB kinase pathways in oncogenic initiation and progression. Oncogene 25(51):6817–6830
31. Jin DY, Chae HZ, Rhee SG, Jeang KT (1997) Regulatory role for a novel human thioredoxin peroxidase in NF-kappaB activation. J Biol Chem 272(49):30952–30961
32. Lo WY, Tsai MH, Tsai Y, Hua CH, Tsai FJ, Huang SY et al (2007) Identification of over-expressed proteins in oral squamous cell carcinoma (OSCC) patients by clinical proteomic analysis. Clin Chim Acta 376(1–2):101–107

33. Nakashima T, Yasumatsu R, Kuratomi Y, Masuda M, Kuwano T, Toh S et al (2006) Role of squamous cell carcinoma antigen 1 expression in the invasive potential of head and neck squamous cell carcinoma. Head Neck 28(1):24–30
34. Imai Y, Sasaki T, Shinagawa Y, Akimoto K, Fujibayashi T (2002) Expression of metastasis suppressor gene (KAI1/CD82) in oral squamous cell carcinoma and its clinico-pathological significance. Oral Oncol 38(6):557–561
35. Do Kyung K, Ahn SG, Park JC, Kanai Y, Endou H, Yoon JH (2004) Expression of L-type amino acid transporter 1 (LAT1) and 4F2 heavy chain (4F2hc) in oral squamous cell carcinoma and its precusor lesions. Anticancer Res 24(3A):1671–1676
36. Schrijvers AH, Gerretsen M, Fritz JM, van Walsum M, Quak JJ, Snow GB et al (1991) Evidence for a role of the monoclonal antibody E48 defined antigen in cell-cell adhesion in squamous epithelia and head and neck squamous cell carcinoma. Exp Cell Res 196(2):264–269
37. Kusukawa J, Ryu F, Kameyama T, Mekada E (2001) Reduced expression of CD9 in oral squamous cell carcinoma: CD9 expression inversely related to high prevalence of lymph node metastasis. J Oral Pathol Med 30(2):73–79
38. Nakashima K, Shimada H, Ochiai T, Kuboshima M, Kuroiwa N, Okazumi S et al (2004) Serological identification of TROP2 by recombinant cDNA expression cloning using sera of patients with esophageal squamous cell carcinoma. Int J Cancer 112(6):1029–1035
39. Jagtap P, McGowan T, Bandhakavi S, Tu ZJ, Seymour S, Griffin TJ et al (2012) Deep metaproteomic analysis of human salivary supernatant. Proteomics 12(7):992–1001
40. Rudney JD, Xie H, Rhodus NL, Ondrey FG, Griffin TJ (2010) A metaproteomic analysis of the human salivary microbiota by three-dimensional peptide fractionation and tandem mass spectrometry. Mol Oral Microbiol 25(1):38–49
41. Kolmeder CA, de Vos WM (2014) Metaproteomics of our microbiome - developing insight in function and activity in man and model systems. J Proteomics 31(97):3–16
42. Heintz-Buschart A, May P, Laczny CC, Lebrun LA, Bellora C, Krishna A et al (2016) Integrated multi-omics of the human gut microbiome in a case study of familial type 1 diabetes. Nat Microbiol 2:16180
43. Broberg M, Doonan J, Mundt F, Denman S, McDonald JE (2018) Integrated multi-omic analysis of host-microbiota interactions in acute oak decline. Microbiome 6(1):21
44. Roy A, Wang S, Meschede-Krasa B, Breffle J, Van Hooser SD (2020) An early phase of instructive plasticity before the typical onset of sensory experience. Nat Commun 11(1):11
45. Washio J, Takahashi N (2016) Metabolomic studies of oral biofilm, oral cancer, and beyond. Int J Mol Sci 2:17(6)
46. Bultman SJ (2014) Emerging roles of the microbiome in cancer. Carcinogenesis 35(2):249–255
47. Wang Y, Zhou Y, Xiao X, Zheng J, Zhou H (2020) Metaproteomics: a strategy to study the taxonomy and functionality of the gut microbiota. J Proteomics 15(219):103737
48. Takahashi N (2015) Oral microbiome metabolism: from "who are they?" to "what are they doing?". J Dent Res 94(12):1628–1637
49. Keyes PH (1968) Research in dental caries. J Am Dent Assoc 76(6):1357–1373
50. Helgeland K (1984) Inhibitory effect of NH4Cl on secretion of collagen in human gingival fibroblasts. Scand J Dent Res 92(5):419–425
51. Kurita-Ochiai T, Seto S, Suzuki N, Yamamoto M, Otsuka K, Abe K et al (2008) Butyric acid induces apoptosis in inflamed fibroblasts. J Dent Res 87(1):51–55
52. Yaegaki K, Qian W, Murata T, Imai T, Sato T, Tanaka T et al (2008) Oral malodorous compound causes apoptosis and genomic DNA damage in human gingival fibroblasts. J Periodontal Res 43(4):391–399
53. Warburg O (1956) On the origin of cancer cells. Science 123(3191):309–314
54. Takahashi N, Washio J, Mayanagi G (2010) Metabolomics of supragingival plaque and oral bacteria. J Dent Res 89(12):1383–1388
55. Ogawa T, Washio J, Takahashi T, Echigo S, Takahashi N (2014) Glucose and glutamine metabolism in oral squamous cell carcinoma: insight from a quantitative metabolomic approach. Oral Surg Oral Med Oral Pathol Oral Radiol 118(2):218–225

56. Humphrey SP, Williamson RT (2001) A review of saliva: normal composition, flow, and function. J Prosthet Dent 85(2):162–169
57. Carpenter GH (2013) The secretion, components, and properties of saliva. Annu Rev Food Sci Technol 4:267–276
58. Cuevas-Cordoba B, Santiago-Garcia J (2014) Saliva: a fluid of study for OMICS. Omics 18(2): 87–97
59. Plottel CS, Blaser MJ (2011) Microbiome and malignancy. Cell Host Microbe 10(4):324–335
60. Larrosa M, Gonzalez-Sarrias A, Garcia-Conesa MT, Tomas-Barberan FA, Espin JC (2006) Urolithins, ellagic acid-derived metabolites produced by human colonic microflora, exhibit estrogenic and antiestrogenic activities. J Agric Food Chem 54(5):1611–1620
61. Mager DL, Haffajee AD, Devlin PM, Norris CM, Posner MR, Goodson JM (2005) The salivary microbiota as a diagnostic indicator of oral cancer: a descriptive, non-randomized study of cancer-free and oral squamous cell carcinoma subjects. J Transl Med 3:27
62. Gong HL, Shi Y, Zhou L, Wu CP, Cao PY, Tao L et al (2013) The composition of microbiome in larynx and the throat biodiversity between laryngeal squamous cell carcinoma patients and control population. PLoS One 8(6):e66476
63. Wei J, Xie G, Zhou Z, Shi P, Qiu Y, Zheng X et al (2011) Salivary metabolite signatures of oral cancer and leukoplakia. Int J Cancer 129(9):2207–2217
64. Sugimoto M, Wong DT, Hirayama A, Soga T, Tomita M (2010) Capillary electrophoresis mass spectrometry-based saliva metabolomics identified oral, breast and pancreatic cancer-specific profiles. Metabolomics 6(1):78–95
65. Gerner EW, Meyskens FL Jr (2004) Polyamines and cancer: old molecules, new understanding. Nat Rev Cancer 4(10):781–792
66. Almadori G, Bussu F, Galli J, Limongelli A, Persichilli S, Zappacosta B et al (2007) Salivary glutathione and uric acid levels in patients with head and neck squamous cell carcinoma. Head Neck 29(7):648–654
67. El-Sayed S, Bezabeh T, Odlum O, Patel R, Ahing S, MacDonald K et al (2002) An ex vivo study exploring the diagnostic potential of 1H magnetic resonance spectroscopy in squamous cell carcinoma of the head and neck region. Head Neck 24(8):766–772
68. Somashekar BS, Kamarajan P, Danciu T, Kapila YL, Chinnaiyan AM, Rajendiran TM et al (2011) Magic angle spinning NMR-based metabolic profiling of head and neck squamous cell carcinoma tissues. J Proteome Res 10(11):5232–5241
69. Tripathi P, Kamarajan P, Somashekar BS, MacKinnon N, Chinnaiyan AM, Kapila YL et al (2012) Delineating metabolic signatures of head and neck squamous cell carcinoma: phospholipase A2, a potential therapeutic target. Int J Biochem Cell Biol 44(11):1852–1861
70. Granato DC, Neves LX, Trino LD, Carnielli CM, Lopes AFB, Yokoo S et al (2021) Meta-omics analysis indicates the saliva microbiome and its proteins associated with the prognosis of oral cancer patients. Biochim Biophys Acta Proteins Proteomics 1869(8):140659
71. Haiser HJ, Turnbaugh PJ (2012) Is it time for a metagenomic basis of therapeutics? Science 336(6086):1253–1255
72. Nicholson JK, Holmes E, Wilson ID (2005) Gut microorganisms, mammalian metabolism and personalized health care. Nat Rev Microbiol 3(5):431–438

Implications and Future Perspectives 13

Shaheerah Khan, Rohita Sinha, Samapika Routray,
and Anshuman Dixit

Abstract

The oral microbiome is considered to be the most diverse in the human body. Around 600–700 bacterial phylotypes have been detected in the oral cavity, most of which belong to the phyla Actinobacteria, Bacteroidetes, Firmicutes, Proteobacteria, Spirochaetes, Synergistetes, and Tenericutes, along with viruses, fungi, protozoa, mycoplasma, and archaea. The human oral cavity is divided into different subsites, including the gingiva, tongue, cheeks, hard and soft palates, hard tissues (teeth), exocrine gland tissue (major and minor salivary glands), supragingival plaque of tooth surfaces, subgingival plaque, maxillary anterior vestibule, and tonsils, all of which are colonized by distinct microbiota with unique features. The composition of the flora in all the subsites is different in healthy individuals and diseased. It is important to understand the composition of the microbiome and the factors that may influence its composition for the development of better diagnostic and therapeutic strategies.

Keywords

Microbial biomarkers · Cancer therapy · Probiotics · Ethnicity · Dietary habits

S. Khan · A. Dixit (✉)
Institute of Life Science, Bhubaneswar, Odisha, India
e-mail: khshaheerah@gmail.com; anshumandixit@gmail.com

R. Sinha
University of Nebraska—Lincoln, Lincoln, NE, USA
e-mail: rohita.sinha@unl.edu

S. Routray
Department of Dentistry, All India Institute of Medical Sciences (AIIMS), Bhubaneswar, Bhubaneswar, Odisha, India
e-mail: dent_samapika@aiimsbhubaneswar.edu.in

S. Routray (ed.), *Microbes and Oral Squamous Cell Carcinoma*,
https://doi.org/10.1007/978-981-19-0592-6_13

13.1 Introduction

The oral microbiome contains a variety of microbes (bacteria, fungi, virus, protozoa) and is one of the most diverse microbiomes in the human body. The alterations in the oral microbiome have the potential to significantly affect other microbiomes e.g., gut microbiome and are associated with predisposition to and the onset of many diseases including cancer. Modulation in the microbiome can be a useful therapeutic approach for various diseases. Therefore, an understanding of the composition of the microbiome, the factors affecting it, and its relationship with physiology and pathology is very important.

13.2 Factors Shaping the Oral Microbiome

13.2.1 Geographical Location/Ethnicity

The diversity in oral microbiome is huge compared to the other microbiome in the human body. Studies have shown that there is a huge diversity in the microflora of different ethnic groups globally. With the evolution of humans from being cave dwellers to living an urban lifestyle, the occupation shifted from hunting, foraging, and farming to modernization and static work-life and changed food habits. All these factors play important roles in the composition of the microbiome. Other environmental factors include diet [1, 2], birth mode (C-section or vaginal birth) [3], and geographical location [4]. Host genetics [5–7] plays a major role in shaping the oral microflora over time. It is not only the microbiome of a healthy oral cavity, but also dental caries and periodontal diseases that differ by ethnicity [8–10].

Studies have shown differences in microflora among the ethnic groups by using different technologies like terminal restriction fragment length polymorphism (t-RFLP) and 16S pyrosequencing. The ethnic groups selected in studies overall included: African, American, African American, European, Asian, Caucasian, Hispanic, non-Hispanic blacks, non-Hispanic whites, Chinese, Indian, Latino, and Burmese. Blekhman et al. did a comparative study between different ethnic groups like African vs. American, African vs. European, African vs. Asian, American vs. European, and American vs. Asian. Li et al. also showed the difference in the saliva of the populations from Alaska, Germany, and Africa. A comparison of the composition of oral microbiome between the healthy populations from major races like the non-Hispanic blacks, non-Hispanic whites, Chinese, and Latinos revealed a significant difference in 33 of 77 genera between these four populations [11].

The most common genera found in all the ethnic groups were *Capnocytophaga, Prevotella, Leptotrichia, Selenomonas, Porphyromonas, Fusobacterium, Corynebacterium, Neisseria, Veillonella,* and TM7-genera-incertae sedis. Phylum Firmicutes was found the highest in Germans and Alaskans, and phylum Proteobacteria in Africans. The common genera between Alaskans and Africans were *Neisseria, Campylobacter, Granulicatella, Megasphaera,* and *Selenomonas,* and between Germans and Africans were *Actinobacillus, Aggregatibacter,* and

Capnocytophaga. Chinese and Latinos showed a higher number of the genera *Filifactor, Staphylococcus, Mycoplasma,* and *Treponema*. The saliva of the Indian population showed enrichment of the family Enterobacteriaceae, Paraprevotellaceae, Spirochaetaceae, phylum Proteobacteria, order Bacteroidales, genus *Streptococcus*, *Solobacterium*, and *Alloprevotella*. Indian state-specific were subspecies of the genus *Chromobacterium* in the Assamese population, Tamil Nadu showed an abundance of genus *Atopobium, Megasphaera*, and *Prevotella,* and Telangana had a higher number of genus *Streptobacillus* and *Bacillus* [12, 13]. In both Burmese and Hispanic population genus *Capnocytophaga* was common. African Americans and Caucasians showed the presence of the genus *Fusobacterium* and *Corynebacterium*, respectively. Some unclassified bacteria of phylum Bacteroidetes, Saccharibacteria and genus *Treponema* were dominant in Burmese; genus *Prevotella* in Caucasians; and genus *Atopobium* and unclassified family Propionibacteriaceae in African American children and Hispanic population [14].

Studies have also shown difference between two other ethnic groups, categorized as the hunter-gatherer group from Uganda and the agricultural group from Congo and Sierra Leone. The difference in the oral microbiome of the hunter-gatherer group and agricultural group was quite prevalent. There is a higher oral microbiome diversity in the hunter-gatherer group than the agricultural group. The hunter-gatherer population showed lower cases of dental caries compared to the agricultural group, probably due to the prevalence of the genus *Haemophilus*, known to maintain the pH and homeostasis in the oral cavity [15].

13.2.2 Dietary Habits

The information on the relationship or link between food, dietary habits, and tastes, to the oral microbiome, is very scarce. The co-evolution between the microbes and humans has existed for long, but there is a change seen in the link and interaction between them [16]. The food ingested has a vital role in shaping the environment of the oral cavity as it changes the pH of the surrounding, thereby affecting the microbiome composition.

Sugar and fatty acids have a significant relation with oral health. The salivary amylase helps to break the carbohydrates (sugar) into the substrate that helps in the growth of oral bacteria, and the bacteria in turn makes the environment more acidic, which is detrimental to the teeth [17]. It has been observed in people with carbohydrate-rich diet a high number of acidogenic and aciduric bacteria, especially genus *Lactobacilli* and *Streptococcus mutans,* in their oral cavity [18, 19], which most of the time are the initial reason for dental caries [20]. The fatty acids (short and long chain, both) on the other hand are also synthesized by the bacteria that help them to survive in the oral cavity, even though the presence of certain fatty acids inhibit their growth [21, 22]. Saturated fatty acids (SFAs) in the diet of people have shown their oral cavity to contain an abundance of phylum Fusobacteria and class Betaproteobacteria. A study showed how a Low Carbohydrate High-Fat (LCHF) diet resulted in a decrease in the number of the genera *Haemophilus*, *Neisseria*, and

Prevotella, whereas an increase in the genus *Streptococcus* [23]. The lipid content of saliva is observed more in the subjects with dental caries. A comparative analysis between diseased state of the oral cavity with periodontitis and gingivitis showed a significantly higher triglycerides and PUFA levels [24]. Other studies show total energy, protein, and fat intake to be positively related with genera *Selenomonas, Ruminococcaceae, Johnsonella*, and *Veillonella*, and negatively correlated with genera *Catonella* and *Peptostreptococcus* [25]. A high-fat/low-fiber diet has proven to promote periodontitis [26], whereas a low-fat/high-fiber diet can be helpful to reduce the disease [27]. Hence, the presence of the bacteria might be the reason or a link between periodontitis and the diet. Similarly, vitamins like vitamin B, C, and E showed more prevalence of the family Leptotrichiaceae, Gemellaceae, Neisseriaceae, and Lachnospiraceae in the oral cavity. There was a positive correlation between glycemic load and the presence of family Lactobacillaceae. Micronutrients, like certain minerals and metals [28, 29], have also been shown to help maintain a balance in the normal habitat of the oral cavity [30]. Though minerals can prove to be beneficial, the elevated amount of certain trace elements like zinc and copper might cause dental caries [31].

Although the pH of saliva is basic, the presence or increase of certain bacteria under the influence of dietary habits may make the environment in the oral cavity acidic. This may catalyse the conversion of nitrite present in the saliva into nitrosamines, which may promote cancer [32].

13.3 Microbial Biomarkers

The presence of a certain community of microbes at a particular site, in a given condition, can act as a marker of that condition. Many studies have shown a difference in the composition of the microbiome, in a normal healthy condition compared to a diseased state (e.g., cancer). A comparative study was done on oral squamous cell carcinoma (OSCC) patient samples shows a major difference in the abundance of the gram-positive and gram-negative populations in oral microbiota. The gram-negative bacteria, phylum Proteobacteria (genus *Escherichia, Vibrio*, and *Salmonella*), and gram-positive bacteria of phylum Firmicutes were the major groups found profoundly associated with OSCC. Other microbial signatures found in OSCC included bacteria—phyla Actinobacteria, Proteobacteria, Firmicutes, and Bacteroidetes; Virus—HPVs (2, 6b, 1, 18, 16, 26, 34), families Poxviridae, Retroviridae, Polyomaviridae, Herpesviridae, Reoviridae, and Orthomyxoviridae; Fungi—genera Fonsecaea, Rhodotorula, Cladophialophora, Geotrichum, Malassezia, Cladosporium, Pleistophora, Pneumocystis, Absidia, and Phialophora; Parasite—genera Hymenolepis, Centrocestus, Dipylidium, Prosthodendrium, Trichinella, Contracaecum, and Toxocara [33]. Another study was done to compare the microbial profiles of healthy individuals, cancer patients, and patients with epithelial precursor lesions, to find the alterations in the abundance of microbial communities in the oral cavity while comparing healthy to cancer patients. There was a significant difference between the transitional epithelial precursor lesion and cancer patients especially in the abundance of genera Bacillus, Enterococcus, Parvimonas, Peptostreptococcus, and Slackia. Such changes in the salivary microbiome can be easily considered as the biomarkers for the diagnosis and prognosis of oral cancer and also of the transition from a tumor lesion to cancer [34].

Studies show that there exists a marker community, whose presence above a certain level signifies a change in the environment, and sometimes can be used to diagnose cancer in its early stage.

13.4 Microbiome and Cancer Therapy

Oral dysbiosis has been seen to influence both cancer progression and therapeutic response. This response is linked with the gut microbiota that can produce metabolites and modify the immune response of the host by increasing the sensitivity of patients to immunotherapy, reducing side effects of chemo and radiotherapy. A patient undergoing cancer treatment might react in accordance with the therapy depending on the microbiome functioning as it modulates the outcome by affecting the efficacy and toxicity [35].

A common condition caused due to chemotherapy and radiotherapy is mucositis. It starts with tissue inflammation which causes ulcers in the oral cavity which again helps in secondary bacterial colonization. The genera *Fusobacterium, Porphyromonas, Treponema*, and *Prevotella* produce cytokines that add to the condition and are shown to change in their numbers during treatment [36]. A significant increase in the number of *Streptococcus mitis* was seen in the patients with nasopharyngeal carcinoma who received radiotherapy. *Streptococcus mitis* breaks down the mucous which helps it to move towards lamina propria causing inflammation [37]. Head and neck cancer patients with a history of undergoing radiation therapy who have shown an increase in *Streptococcus anginosus, Streptococcus mitis, Streptococcus mutans, Streptococcus oralis, Streptococcus sanguinis*, and *Streptococcus sobrinus*, develop oral mucositis. Xerostomia is another condition besides oral mucositis that alters the biofilm composition of the oral cavity. This not only reduces the blood supply but also reduces immune response causing a delay in repair as well. Xerostomia shows an abundance of bacteria from the family Enterobacteriaceae and the presence of fimbriae facilitates its adhesion to the tumor cells by increased proteolytic activity on fibronectin in the oral cavity through adhesion receptors, enhancing not only its colonization but also inflammation through IL-1 and IL-6 [38].

There have been studies that show the beneficial effects of the microbiome in enhancing cancer therapy. *F. nucleatum* with *Bacteroides fragilis* and *Escherichia coli* improved the survival of patients in an Adoptive Cell Therapy (ACT) by increasing the production of cytokine and T cell infiltration [39]. Another study showed the effect of *Lactobacillus fermentum* in stimulating the immune response in patients who have undergone CpG-oligonucleotide immunotherapy [40]. Studies done in the 1800s have shown that the infestation of certain bacterial species helps in the reduction of cancer [41]. *Streptococcus pyogenes* has been shown to shrink the size of a malignant tumor, initially which was given in immunotherapy. The cons of using bacteria to treat cancer include short-term benefits and the recurrence, applying to the fact that therapies like chemo- and radiotherapy showed better results. However, chemotherapy and radiotherapy have their own side effects, hence to combat that, "bacteria-mediated tumor therapy" could be an attractive alternative [42, 43]. The first step is to identify the species having the least toxicity and pathogenicity to the host, and able to replicate in the tumor microenvironment, followed by the systemic administration of the bacteria in the host. The genera *Clostridium, Bifidobacterium, E. coli*, and attenuated *Salmonella* have been used in bacteria-mediated tumor therapy [44, 45]. The advantage of this therapy is that it can penetrate solid tumors and is inexpensive. Another approach is called "combinatorial therapy", which involves microbes-mediated therapies combined with other kinds of cancer therapies. Here a photo-thermal agent, such as melanin-like poly-dopamine (pDA), coated with a genetically stable S. typhi strain (VNP20009), targets the hypoxic and necrotic tumor area. A mouse model of the tumor was irradiated with a near-infrared laser that caused a targeted elimination of the tumor, without metastasis or relapse [46]. However, studies done are less, but these techniques might have the potential to treat OSCC.

13.5 Association of Noncoding RNAs with the Microbiome and Oral Cancer

MicroRNAs are a class of noncoding RNAs of 20–22 nucleotides in length and have been linked with the process of driving cancer. They are also known to have a relation with the microbiota [47–49]. Studies have shown that certain miRNAs have the ability to penetrate the bacteria and modulate its gene expression altering the composition of the microbiota and affecting its growth [50–53]. This can be easily adopted as a therapeutic approach towards restoring the eubiotic community, by specifically targeting the pathogenic bacteria and reducing their growth. This interrelationship also gives the ability to the microbes to alter the oral miRNAs, by the production of different metabolites that leads to significant changes in the metabolism of the host cells [54–56].

13.6 Probiotics and Oral Cancer

The use of probiotics to improve health has received a lot of attention in recent years [57]. It has been demonstrated that probiotics help to improve the immune system and overall health. These probiotics can maintain a healthy balance of intestinal flora and their metabolic activity. They may also secrete metabolites, e.g., short-chain fatty acids (SCFAs) and lactic acid, that can stimulate the immune system and have antitumor properties [58]. The lactobacillus in the gut has been reported affecting regression of carcinogenesis through immunomodulation. Probiotics have been shown to modulate the production of anti-inflammatory cytokines that may further help in the prevention of cancers [59]. Probiotics may be effective in preventing and treating cancer therapy-induced oral mucositis [60]. Recently, it has been reported that probiotics with potential therapeutic effects against oral carcinogenesis include *Acetobacter syzygii, AJ2, Lactobacillus plantarum,* and *Lactobacillus salivarius REN*. Among them, *L. salivarius REN* was reported to reduce the risk of development of oral cancer significantly [61]. However, more studies are required to establish probiotics as a therapeutic option for oral cancer.

13.7 Conclusion

Microbiome in itself is a huge world that requires greater exploration to find out their relevance in human health and disease. More studies based upon omics technologies are required to understand the inhabitants of a microbiome better. The role of microbes in causing disease has been studied extensively, but exploring the side where microbes can be used to treat the diseases such as cancer could be very interesting to explore. The fact that microbiome has been explored for its usage in cancer therapeutics is itself a novel achievement. Cancer immunotherapy and usage of noncoding RNAs and probiotics are a few examples that restore back the faith that these microbes can be utilized well to reverse the ill effects caused by other microbes. A macroscopic philosophy that comes from the microscopic world, *"where there is bad, there is also good,"* projects towards the future that holds many more such therapeutics to be unveiled, and researchers are hopeful in finding new and sustainable ways for diagnostics and therapeutics discovery. The study of the microbial omics of OSCC, by bringing the knowledge of the detailed role of microbes in oral cancer, is a step taken towards the solution.

References

1. Kato I, Vasquez A, Moyerbrailean G, Land S, Djuric Z, Sun J et al (2017) Nutritional correlates of human oral microbiome. J Am Coll Nutr 36(2):88–98
2. Adler CJ, Dobney K, Weyrich LS, Kaidonis J, Walker AW, Haak W et al (2013) Sequencing ancient calcified dental plaque shows changes in oral microbiota with dietary shifts of the Neolithic and Industrial revolutions. Nat Genet 45(4):450–455, 5e1
3. Lif Holgerson P, Harnevik L, Hernell O, Tanner AC, Johansson I (2011) Mode of birth delivery affects oral microbiota in infants. J Dent Res 90(10):1183–1188
4. Li J, Quinque D, Horz HP, Li M, Rzhetskaya M, Raff JA et al (2014) Comparative analysis of the human saliva microbiome from different climate zones: Alaska, Germany, and Africa. BMC Microbiol 14:316
5. Gomez A, Espinoza JL, Harkins DM, Leong P, Saffery R, Bockmann M et al (2017) Host genetic control of the oral microbiome in health and disease. Cell Host Microbe 22(3): 269–278, e3
6. Corby PM, Bretz WA, Hart TC, Filho MM, Oliveira B, Vanyukov M (2005) Mutans streptococci in preschool twins. Arch Oral Biol 50(3):347–351
7. Corby PM, Bretz WA, Hart TC, Schork NJ, Wessel J, Lyons-Weiler J et al (2007) Heritability of oral microbial species in caries-active and caries-free twins. Twin Res Hum Genet 10(6): 821–828. https://doi.org/10.1375/twin.10.6.821

8. Golden SH, Brown A, Cauley JA, Chin MH, Gary-Webb TL, Kim C et al (2012) Health disparities in endocrine disorders: biological, clinical, and nonclinical factors—an Endocrine Society scientific statement. J Clin Endocrinol Metab 97(9):E1579–E1639
9. Naorungroj S, Slade GD, Divaris K, Heiss G, Offenbacher S, Beck JD (2017) Racial differences in periodontal disease and 10-year self-reported tooth loss among late middle-aged and older adults: the dental ARIC study. J Public Health Dent 77(4):372–382
10. Schork NJ (2015) Personalized medicine: time for one-person trials. Nature 520(7549):609–611
11. Mason MR, Nagaraja HN, Camerlengo T, Joshi V, Kumar PS (2013) Deep sequencing identifies ethnicity-specific bacterial signatures in the oral microbiome. PLoS One 8(10):e77287
12. Acharya A, Chan Y, Kheur S, Kheur M, Gopalakrishnan D, Watt RM et al (2017) Salivary microbiome of an urban Indian cohort and patterns linked to subclinical inflammation. Oral Dis 23(7):926–940
13. Sarkar A, Stoneking M, Nandineni MR (2017) Unraveling the human salivary microbiome diversity in Indian populations. PLoS One 12(9):e0184515
14. Premaraj TS, Vella R, Chung J, Lin Q, Hunter P, Underwood K et al (2020) Ethnic variation of oral microbiota in children. Sci Rep 10(1):14788
15. Gupta VK, Paul S, Dutta C (2017) Geography, ethnicity or subsistence-specific variations in human microbiome composition and diversity. Front Microbiol 8:1162
16. Sun J, Chang EB (2014) Exploring gut microbes in human health and disease: pushing the envelope. Genes Dis 1(2):132–139
17. Touger-Decker R, van Loveren C (2003) Sugars and dental caries. Am J Clin Nutr 78(4):881S–892S
18. Wade WG (2013) The oral microbiome in health and disease. Pharmacol Res 69(1):137–143
19. Hojo K, Nagaoka S, Ohshima T, Maeda N (2009) Bacterial interactions in dental biofilm development. J Dent Res 88(11):982–990
20. Gross EL, Leys EJ, Gasparovich SR, Firestone ND, Schwartzbaum JA, Janies DA et al (2010) Bacterial 16S sequence analysis of severe caries in young permanent teeth. J Clin Microbiol 48(11):4121–4128
21. Huang CB, George B, Ebersole JL (2010) Antimicrobial activity of n-6, n-7 and n-9 fatty acids and their esters for oral microorganisms. Arch Oral Biol 55(8):555–560
22. Huang CB, Alimova Y, Myers TM, Ebersole JL (2011) Short- and medium-chain fatty acids exhibit antimicrobial activity for oral microorganisms. Arch Oral Biol 56(7):650–654
23. Murtaza N, Burke LM, Vlahovich N, Charlesson B, O'Neill HM, Ross ML et al (2019) Analysis of the effects of dietary pattern on the oral microbiome of elite endurance athletes. Nutrients 13:11(3)
24. Figueredo CM, Martinez GL, Koury JC, Fischer RG, Gustafsson A (2013) Serum levels of long-chain polyunsaturated fatty acids in patients with periodontal disease. J Periodontol 84(5): 675–682
25. Cattaneo C, Riso P, Laureati M, Gargari G, Pagliarini E (2019) Exploring associations between Interindividual differences in taste perception, oral microbiota composition, and reported food intake. Nutrients 24:11(5)
26. Tomofuji T, Furuta M, Ekuni D, Irie K, Azuma T, Iwasaki Y et al (2011) Relationships between eating habits and periodontal condition in university students. J Periodontol 82(12):1642–1649
27. Kondo K, Ishikado A, Morino K, Nishio Y, Ugi S, Kajiwara S et al (2014) A high-fiber, low-fat diet improves periodontal disease markers in high-risk subjects: a pilot study. Nutr Res 34(6): 491–498
28. Garhammer P, Hiller KA, Reitinger T, Schmalz G (2004 Dec) Metal content of saliva of patients with and without metal restorations. Clin Oral Investig 8(4):238–242
29. Olmedo P, Pla A, Hernandez AF, Lopez-Guarnido O, Rodrigo L, Gil F (2010) Validation of a method to quantify chromium, cadmium, manganese, nickel and lead in human whole blood, urine, saliva and hair samples by electrothermal atomic absorption spectrometry. Anal Chim Acta 659(1–2):60–67

30. Garcia-Godoy F, Hicks MJ (2008) Maintaining the integrity of the enamel surface: the role of dental biofilm, saliva and preventive agents in enamel demineralization and remineralization. J Am Dent Assoc 139(Suppl):25S–34S
31. Hegde MN, Hegde ND, Ashok A, Shetty S (2014) Biochemical indicators of dental caries in saliva: an in vivo study. Caries Res 48(2):170–173
32. Combet E, Paterson S, Iijima K, Winter J, Mullen W, Crozier A et al (2007) Fat transforms ascorbic acid from inhibiting to promoting acid-catalysed N-nitrosation. Gut 56(12):1678–1684
33. Banerjee S, Tian T, Wei Z, Peck KN, Shih N, Chalian AA et al (2017) Microbial signatures associated with oropharyngeal and oral squamous cell carcinomas. Sci Rep 7(1):4036
34. Lee WH, Chen HM, Yang SF, Liang C, Peng CY, Lin FM et al (2017) Bacterial alterations in salivary microbiota and their association in oral cancer. Sci Rep 7(1):16540
35. Panebianco C, Andriulli A, Pazienza V (2018) Pharmacomicrobiomics: exploiting the drug-microbiota interactions in anticancer therapies. Microbiome 6(1):92
36. Hou J, Zheng H, Li P, Liu H, Zhou H, Yang X (2018) Distinct shifts in the oral microbiota are associated with the progression and aggravation of mucositis during radiotherapy. Radiother Oncol 129(1):44–51
37. Zhu XX, Yang XJ, Chao YL, Zheng HM, Sheng HF, Liu HY et al (2017) The potential effect of oral microbiota in the prediction of mucositis during radiotherapy for nasopharyngeal carcinoma. EBioMedicine 18:23–31
38. Gaetti-Jardim E Jr, Jardim ECG, Schweitzer CM, da Silva JCL, Oliveira MM, Masocatto DC et al (2018) Supragingival and subgingival microbiota from patients with poor oral hygiene submitted to radiotherapy for head and neck cancer treatment. Arch Oral Biol 90:45–52
39. Cremonesi E, Governa V, Garzon JFG, Mele V, Amicarella F, Muraro MG et al (2018) Gut microbiota modulate T cell trafficking into human colorectal cancer. Gut 67(11):1984–1994
40. Iida N, Dzutsev A, Stewart CA, Smith L, Bouladoux N, Weingarten RA et al (2013) Commensal bacteria control cancer response to therapy by modulating the tumor microenvironment. Science 342(6161):967–970
41. Coley WB II (1891) Contribution to the knowledge of sarcoma. Ann Surg 14(3):199–220
42. Mengesha A, Dubois L, Chiu RK, Paesmans K, Wouters BG, Lambin P et al (2007) Potential and limitations of bacterial-mediated cancer therapy. Front Biosci 1(12):3880–3891
43. Nallar SC, Xu DQ, Kalvakolanu DV (2017) Bacteria and genetically modified bacteria as cancer therapeutics: current advances and challenges. Cytokine 89:160–172
44. Parker RC, Plummer HC et al (1947) Effect of histolyticus infection and toxin on transplantable mouse tumors. Proc Soc Exp Biol Med 66(2):461–467
45. Yazawa K, Fujimori M, Amano J, Kano Y, Taniguchi S (2000) Bifidobacterium longum as a delivery system for cancer gene therapy: selective localization and growth in hypoxic tumors. Cancer Gene Ther 7(2):269–274
46. Chen W, Wang Y, Qin M, Zhang X, Zhang Z, Sun X et al (2018) Bacteria-driven hypoxia targeting for combined biotherapy and photothermal therapy. ACS Nano 12(6):5995–6005
47. Dalmasso G, Nguyen HT, Yan Y, Laroui H, Charania MA, Ayyadurai S et al (2011) Microbiota modulate host gene expression via microRNAs. PLoS One 6(4):e19293
48. Peck BC, Mah AT, Pitman WA, Ding S, Lund PK, Sethupathy P (2017) Functional transcriptomics in diverse intestinal epithelial cell types reveals robust microRNA sensitivity in intestinal stem cells to microbial status. J Biol Chem 292(7):2586–2600
49. Yuan C, Subramanian S (2019) microRNA-mediated tumor-microbiota metabolic interactions in colorectal cancer. DNA Cell Biol 38(4):281–285
50. Golan-Gerstl R, Elbaum Shiff Y, Moshayoff V, Schecter D, Leshkowitz D, Reif S (2017) Characterization and biological function of milk-derived miRNAs. Mol Nutr Food Res 61(10)
51. Liu S, da Cunha AP, Rezende RM, Cialic R, Wei Z, Bry L et al (2016) The host shapes the gut microbiota via fecal microRNA. Cell Host Microbe 19(1):32–43
52. Teodori L, Petrignani I, Giuliani A, Prattichizzo F, Gurau F, Matacchione G et al (2019) Inflamm-aging microRNAs may integrate signals from food and gut microbiota by modulating common signalling pathways. Mech Ageing Dev 182:111127

53. Zhang L, Chen T, Yin Y, Zhang CY, Zhang YL (2019) Dietary microRNA—a novel functional component of food. Adv Nutr 10(4):711–721
54. Masotti A (2012) Interplays between gut microbiota and gene expression regulation by miRNAs. Front Cell Infect Microbiol 2:137
55. Miro-Blanch J, Yanes O (2019) Epigenetic regulation at the interplay between gut microbiota and host metabolism. Front Genet 10:638
56. Patrignani P, Tacconelli S, Bruno A (2014) Gut microbiota, host gene expression, and aging. J Clin Gastroenterol 48(Suppl 1):S28–S31
57. Lu K, Dong S, Wu X, Jin R, Chen H (2021) Probiotics in cancer. Front Oncol 11:638148
58. Legesse Bedada T, Feto TK, Awoke KS, Garedew AD, Yifat FT, Birri DJ (2020) Probiotics for cancer alternative prevention and treatment. Biomed Pharmacother 129:110409
59. Hill C, Guarner F, Reid G, Gibson GR, Merenstein DJ, Pot B et al (2014) Expert consensus document. The International Scientific Association for Probiotics and Prebiotics consensus statement on the scope and appropriate use of the term probiotic. Nat Rev Gastroenterol Hepatol 11(8):506–514
60. Shu Z, Li P, Yu B, Huang S, Chen Y (2020) The effectiveness of probiotics in prevention and treatment of cancer therapy-induced oral mucositis: a systematic review and meta-analysis. Oral Oncol 102:104559
61. Wan Mohd Kamaluddin WNF, Rismayuddin NAR, Ismail AF, Mohamad Aidid E, Othman N, Mohamad NAH et al (2020) Probiotic inhibits oral carcinogenesis: a systematic review and meta-analysis. Arch Oral Biol 118:104855

The manufacturer's authorised representative in the EU is Springer Nature Customer Service Centre GmbH, Europaplatz 3, 69115 Heidelberg, Germany. If you have any concerns regarding our products, please contact ProductSafety@springernature.com

Printed and bound by CPI Group (UK) Ltd, Croydon, CR0 4YY
30/09/2026
02225265-0014